▶ The Pathology of Communicative Capitalism

DOI: 10.1057/9781137394781.0001

Other Palgrave Pivot titles

Philip Whitehead: Reconceptualising the Moral Economy of Criminal Justice: A New Perspective

Thomas Kaiserfeld: Beyond Innovation: Technology, Institution and Change as Categories for Social Analysis

Dirk Jacob Wolfson: The Political Economy of Sustainable Development: Valuation, Distribution, Governance

Twyla J. Hill: Family Caregiving in Aging Populations

Alexander M. Stoner and Andony Melathopoulos: Freedom in the Anthropocene: Twentieth Century Helplessness in the Face of Climate Change

Christine J. Hong: Identity, Youth, and Gender in the Korean American Christian Church

Cenap Çakmak and Murat Ustaoğlu: Post-Conflict Syrian State and Nation Building: Economic and Political Development

Richard J. Arend: Wicked Entrepreneurship: Defining the Basics of Entreponerology

Rubén Arcos and Randolph H. Pherson (editors): Intelligence Communication in the Digital Era: Transforming Security, Defence and Business

Jane Chapman, Dan Ellin and Adam Sherif: Comics, the Holocaust and Hiroshima

AKM Ahsan Ullah, Mallik Akram Hossain and Kazi Maruful Islam: Migration and Worker Fatalities Abroad

Debra Reddin van Tuyll, Nancy McKenzie Dupont and Joseph R. Hayden: Journalism in the Fallen Confederacy

Michael Gardiner: Time, Action and the Scottish Independence Referendum

Tom Bristow: The Anthropocene Lyric: An Affective Geography of Poetry, Person, Place

Shepard Masocha: Asylum Seekers, Social Work and Racism

Michael Huxley: The Dancer's World, 1920–1945: Modern Dancers and Their Practices Reconsidered

Michael Longo and Philomena Murray: Europe's Legitimacy Crisis: From Causes to Solutions

Mark Lauchs, Andy Bain and Peter Bell: Outlaw Motorcycle Gangs: A Theoretical Perspective

Majid Yar: Crime and the Imaginary of Disaster: Post-Apocalyptic Fictions and the Crisis of Social Order

Sharon Hayes and Samantha Jeffries: Romantic Terrorism: An Auto-Ethnography of Domestic Violence, Victimization and Survival

Gideon Maas and Paul Jones: Systemic Entrepreneurship: Contemporary Issues and Case Studies

Surja Datta and Neil Oschlag-Michael: Understanding and Managing IT Outsourcing: A Partnership Approach

Keiichi Kubota and Hitoshi Takehara: Reform and Price Discovery at the Tokyo Stock Exchange: From 1990 to 2012

DOI: 10.1057/9781137394781.0001

palgrave▸pivot

The Pathology of Communicative Capitalism

David W. Hill
University of Liverpool, UK

palgrave
macmillan

DOI: 10.1057/9781137394781.0001

First published 2015 by
PALGRAVE MACMILLAN

Palgrave Macmillan in the UK is an imprint of Macmillan Publishers Limited, registered in England, company number 785998, of Houndmills, Basingstoke, Hampshire RG21 6XS.

Palgrave Macmillan in the US is a division of St Martin's Press LLC, 175 Fifth Avenue, New York, NY 10010.

Palgrave Macmillan is the global academic imprint of the above companies and has companies and representatives throughout the world.

Palgrave® and Macmillan® are registered trademarks in the United States, the United Kingdom, Europe and other countries.

ISBN: 978–1–137–39479–8 EPUB
ISBN: 978–1–137–39478–1 PDF
ISBN: 978–1–137–39477–4 Hardback

A catalogue record for this book is available from the British Library.

A catalog record for this book is available from the Library of Congress.

www.palgrave.com/pivot

DOI: 10.1057/9781137394781

▶ *Dedicated to the memory of Kenneth Morelli (1930–2014) and to the enduring spirit of his beautiful wife, Muriel.*

DOI: 10.1057/9781137394781.0001

Contents

Introduction   1

1 Cognitive Labour   12

2 Communicative Disease   29

3 Social Anxiety   40

4 Pathological Development   55

References   67

Index   73

DOI: 10.1057/9781137394781.0001

Introduction

Abstract: *This chapter sets out the basic structure of the book. It also introduces the key ideas that underpin it: the productive role of communication in generating value in the economy; the disappointment that technological developments have not led to a reduction in work time; the triumph of automation over worker autonomy; and the need to reevaluate the role of networked technologies in society in order to revalue communication itself. This chapter also introduces some of the core ideas of the Autonomist theory of labour and Jodi Dean's definition of our present socioeconomic system as* communicative capitalism.

Hill, David W. *The Pathology of Communicative Capitalism*. Basingstoke: Palgrave Macmillan, 2015. DOI: 10.1057/9781137394781.0002.

In 2015 the technology magazine *Wired* reported the launch of a new Facebook service, Facebook at Work (see Alba 2015). The magazine ran with the headline 'Facebook at Work Launches So You Can Never Not Be On Facebook'. Strict grammarians and the Microsoft Word spell-checker may balk at 'never not' but 'never not on' summarises the ethos not only of the social networking corporation in question, but also the socio-economic constellation of communicative capitalism. Facebook, Inc. has noticed – because they have access to this sort of data – that people are using their social networking site during working hours and so have developed a portal that seeks to take advantage of this. It works much like the normal Facebook site but allows workers to connect with their colleagues, who are not necessarily part of the user's friendship group on their social network. By creating a new portal, one that has a different design and colour scheme, the idea is that people who might feel reticent to be caught socialising during work hours can use Facebook in the office, since their supervisor will be able to recognise that this is the work tool and not the regular procrastination service, and so Facebook can capture more of the at work market. At its heart this is a move designed to capture a greater share of our screen time, monetising our attention in a new market since user uptake on the site is slowing (see Alba 2015). Lars Rasmussen, Facebook director of engineering, sells this new service as a way of streamlining communication in the workplace: 'We have found that using Facebook as a work tool makes our work day more efficient' (in Alba 2015). Workers will be able to see posts from their colleagues and chat through the instant messaging facility, much like with the social version but oriented towards more profitable ends. Rasmussen argues that not only is this more efficient in terms of productivity, but that it will also allow workers to develop a more stable work-life balance: 'Some people are less comfortable than others using their personal Facebook in the work context... With Facebook at Work, you get the option of completely separating the two' (in Alba 2015). This is a somewhat fanciful claim. The launch of Facebook at Work is an acknowledgement of a transformation in labour, where the communicational has become a core productive force. At the same time, it tacitly acknowledges the role of communication technologies in workplace surveillance; the image of the boss peering over one's shoulder to check what is taking place on the work terminal is used as motivation to migrate to a tool that would allow corporations to have real-time oversight of users' communicational content. Users are being encouraged to place their communication into

DOI: 10.1057/9781137394781.0002

an enclosure of control constituted by code. Such a development does not separate the social from the productive but rather is indicative of the way that workers' social capacities – personality, personability, the compulsion for interaction – can be expropriated and exploited in ways that are easily monitored. These skills have become valuable resources, and Facebook at Work would seek to translate our enthusiasm for socialising and staying connected through social networking sites into an enthusiasm for work. If and when such a product becomes widely adopted, it is difficult to imagine that it will not follow the path of the work email, seeping out from the confines of the traditional working day, becoming something that is accessed and updated in leisure time, workers' willingness (conditioning) to labour outside of work time even more easy to monitor than with emails. With plans to develop Facebook at Work apps for smartphones and tablets this becomes all the more likely, since these mobile devices have become pocket work sirens, maintaining the connection to work at all hours of the day. Facebook, Inc. is not only tapping into a new market but catching up with a complete transformation in the world of work.

The aim of this book is to chart the ways in a which a system of capital that operates on the value of cognitive labour and the productivity of communication, captured by new communication technologies, creates precarious conditions in society, from the instability of flexible work to mental fragilities wrought by constant attentive stress, the social fragmentation of communication valued by its exchange efficiency rather than its capacity to bond, and, ultimately, the erosion of solidarity from which any kind of resistance could be launched. But first, it is useful to ask how we got here in the first place, how the technological developments that ought to have freed workers from the excessive demands of capital, the technologies that promised a brighter future, have instead redoubled the burden. The Autonomist theorist Franco Berardi provides a brief history of the future, of the way that the radical potential for technology to transform work conditions has been neutralised by capital, such that development can no longer be understood as leading to a better tomorrow. Autonomists emphasise the creativity of work and the desirability of worker autonomy from the power of capital (see Gill and Pratt 2008: 5–9; Lotringer 2007: v–xvi; Wright 2002). A key site of struggle, both conceptually and for activism, was centred around the Fiat car factory in Turin, where downsizing and automation in the late 1970s began what Marco Revelli (1996: 116) calls 'a process of mobility that

DOI: 10.1057/9781137394781.0002

neutralised the factory as a place of belonging and aggregation, and sent individuals back to a state of atomization and isolation' (see also Pansa 2007: 24–27). The defeat of this worker movement, who had been calling for technological productivity to initiate a reduction in work time and freedom from the sadness of labour, signalled the victory of the speed and mobility of capital. For Christian Marazzi (2011b: 30), the 1970s transformed the world of work, bringing with it 'reduction in the cost of labor, attacks on unions, automisation and robotisation of entire labour processes, delocalisation to countries with low wages, precarization of labour and diversification of consumption models'. Just as western societies had begun to reach the point at which work could be progressively scaled back, increasing leisure time and enjoyment, capital flexed and completely changed the rules of the game. According to Berardi (2009a: 18), the strikes and protests in Italy were not concerned with labour parties or socialist ideologies, but by the simple demand for less work, less misery, by making the most of the new technological horizon; its defeat shows that even the more modest political challenges to capitalist rule are seen as a deadly threat to the system, and the quality of life as something to be sacrificed to capital. Berardi (2009a: 27–28) picks out a series of events in 1977 that he suggests cements its place in history as the year the future died. This was the year that the *Rote Armee Fraktion* engaged in its German Autumn of kidnappings, hijackings and assassinations, culminating in the supposed suicides of the Baader-Meinhof captives. At the same time, Jean-François Lyotard was writing *The Postmodern Condition*, his essay on the role of technology in knowledge production, coinciding with the *RAF* deaths as if the end of radicalism was first required before the transformation of knowledge by techno-scientific capitalism – and ultimately the transformation of the human – could fully take hold. The end of the year saw the release of *Saturday Night Fever*, the John Travolta dance movie, which depicted 'a new working class, happy to be exploited all week long in exchange for some fun in the disco' (Berardi 2009a: 27). And it was the year that Steve Jobs, Ronald Wayne and Steve Wozniak incorporated Apple Computer, Inc.

Mark Fisher (2009: 13) argues that, unlike fascism or Stalinism, capitalism does not need propaganda since it 'can proceed perfectly well, in some ways better, without anyone making a case for it'. In part this seems correct, since the system works best when we all assume that it is in fact perfectly natural; there is no need to make a case for something when the desired aim is to ensure that people forget there was ever an

DOI: 10.1057/9781137394781.0002

alternative. However, a different kind of propaganda is at play, one that seeks not to make a case for capitalism in any direct way, but instead indirectly by creating an enthusiasm, even fetishism, for new technologies and communication networks that are not only product samples for a consumer society, but the means by which the system sustains and expands itself. The product launches for new Apple products replicate the spectacle of *Saturday Night Fever*, selling the idea that a month of exploitation at work is a fair price to pay for some fun with a brand new notebook, tablet or smartphone. Enthusiasm for communication technologies too easily becomes an enthusiasm for the immaterial labour and frictionless exchange of information that marks the new capitalism begun in the 1970s. Enthusiastic political activists, seduced by the potential for new connections and sites of struggle through social networking groups, micro-blogging, online petitions and so on, celebrate the sought-after hardware that sustains these activities, 'unaware that their message is indistinguishable from Apple's' (Dean 2009: 9). Our energies are diverted into the architecture of the system that exploits us. We are encouraged to participate in information, to be available to communication, at the same time that these become profitable states in both work and leisure. We desire faster connection times and greater processing speeds, at the same time that the demand to save time becomes the operating principle of the system of capital. 'Speed', writes Paul Virilio (2012: 38), 'the cult of speed, is the propaganda of progress'. This is not the progress of the Enlightenment or humanism, but one constituted entirely by acceleration. This is a progress that has reverted to superstition and blind-faith:

> What we are living through now has taken the shape of a religion; it is not unlike a return to sun worship where speed has replaced light. We are experiencing the return of a major myth supported by the propaganda of progress...we are no longer in the Enlightenment: we are in the century of light speed. (Virilio 2012: 41)

But where religions were invented to explain natural phenomena such as earthquakes or volcanoes as the anger of a god, such that appeasement might lead to stability, we are here worshipping something that is destabilising society without any concern for the consequences, such that our observance only makes things worse. The mobility of capital, lack of belonging and aggregation at work, atomisation and isolation, increasing rationalisation, the vulnerability of leisure time, enclosures of control and general precariousness are all a result of this cult of speed, buying

DOI: 10.1057/9781137394781.0002

into a kind of progress that is little more than inhuman development. Facebook at Work is simply one example amongst many of acceleration and flexibility as operational principles.

Today, the future is a myth born of communicative capitalism and its unceasing bid for expansion. This is not the radical future of revolution and the institution of new social conditions, but simply an expectation that the movement of time will bring more of the same, only faster and more totalising. The brief interruption by cyberculture towards the end of the last century constituted a belief that things could be different, a generation who trusted in the future, who believed it could be better than the present, but whose cyberculture has now given way 'to the imagination of the global mind, hyperconnected and infinitely powerful' (Berardi 2011: 17). The future is no longer on the side of the revolutionary. The radical potential of new technologies has been co-opted and neutralised to achieve the neoliberalisation of society and economy. The idea that development would mean an expansion of free time, emancipation from the confines of labour, has been grotesquely turned against workers, as capital accrues the freedom and time is enslaved. We have seen a three-fold abstraction of society: the immaterialisation of labour, as when activities that were not previously considered work, such as communication, become economically valuable; the digitalisation of communication, in both labour and leisure; and the informationalisation of a financial system that has seemingly shaken free of any human reins. This economy was never so much concerned with knowledge but with moulding subjectivity; we acquiesced in the rhetoric of an information society which was little more than the glorification of an informational financial system that occupies society and its citizens with debt management (Lazzarato 2014: 9-10). Weakened by debt, society becomes all the more pliable, such that populations will submit to modes of labour, consumption and communication that are most profitable.

With a post-Fordist work environment, where labour becomes communication, we see the inseparability of work and life:

> The Fordist factory was crudely divided into blue and white collar work, with the different types of labor physically delimited by the structure of the building itself. Laboring in noisy environments, watched over by managers and supervisors, workers had access to language only in their breaks, in the toilet, at the end of the working day, or when they were engaged in sabotage, because communication interrupted production. But in post-Fordism, people work by communicating. (Fisher 2009: 34)

DOI: 10.1057/9781137394781.0002

This shift enmeshes labour with sociality, re-shaping social subjectivity so that it can be captured for profit, constituting the present epochal moment when, as Yann Moulier Boutang (2011: 57) suggests, the Internet and the computer become emblematic in the same way that the coal mine or the steam engine were to industrial capitalism. Communication and production are now one and the same; this necessitates that we submit our new technological forms of communication to critique, since we can no longer assume that they are neutral, as Berardi cautions:

> The machine pretends to be neutral, purely mathematical, but we know that its procedures are only the technical reification of social interests: profit, accumulation, competition – these are the criteria underlying the automatic procedures embedded in the machine. (2011: 58)

The title of this book takes as its key site of inquiry what might be called communicative capitalism. This is a term taken from the work of Jodi Dean (see 2009; 2010; 2012) who uses it to pick out a system character-ised by the 'amplified role of communication in production' (2012: 18). Communicative capitalism is a mode of production that exerts control over society by conditioning its occupants towards its own ends, namely the acceleration of its informational regime through establishing the primacy of cognitive labour and consumptive modes of communica-tion that generate value. In its ideal form this is a state in which work-ers toil down the data mine under ever more precarious conditions, whilst in their free time they are subjected to data-mining through the communicational technologies with which they attempt to find respite and social connection. Facebook at Work provides one example of this exploitation of the overproduction of information and its communica-tion, but as Maurizio Lazzarato (2014: 37) notes, we can observe similar of Google, which operates as a databank for marketing, gathering information about our web navigation, purchases, cultural tastes and how we like to spend our leisure time, generating profiles that 'are mere relays of inputs and outputs in production-consumption machines'. Communicative capitalism constitutes an enclosure that entraps and exploits every facet of our existence. To get the future back on track we have to scrutinise the communication technologies that we too readily fetishise, to question how we are being constrained into relations and social organisations that are not necessarily in our own interests, and to slow down long enough to question the mythology of speed, accelera-tion and interactivity. It is time to stop eulogising the advances and to

DOI: 10.1057/9781137394781.0002

question the reorganisation of society by immateriality and the ideas that support it – and yet in many ways the change we might want to communicate is stuck in the circuits of the system that needs changing.

Dean (2009: 2) defines communicative capitalism more expansively as 'the materialization of ideals of inclusion and participation in information, entertainment, and communication technologies in ways that capture resistance and intensify global capitalism'. The exploitation of communication is reliant on exploiting our enjoyment of the system, of connecting, networking, consuming information and cultural commodities, capturing our social desires in networks of production and surveillance (Dean 2010: 4). By offering something for everyone it includes rather than excludes, whilst it fetishises communication in order to ensure participation:

> It embeds us in a mindset wherein the number of friends one has on Facebook or Myspace, the number of page-hits one gets on one's blogs, and the number of videos featured on one's YouTube channel are the key markers of success, and details such as duration, depth of commitment, corporate and financial influence, access to structures of decision making, and the narrowing of political struggle to the standards of do-it-yourself entertainment culture become the boring preoccupations of baby-boomers stuck in the past. (Dean 2009: 17)

Far from questioning power or exploitation users become receptive to networking growth, web traffic and transactional data in the way one might expect of someone with the 'bullshit' (Graeber 2013) job title of something like *Social Media Marketing Executive*. All of this information – about how far content circulates, as with retweets on Twitter or re-blogging, the scale of social networks, consumptive behaviours – become part of a user's profile, an expression of identity or perhaps of social success. It is, at the same time, transparent evidence of the user's unrecompensed productivity and of their being tracked and traced, measured and valued. We have what Jean-François Lyotard once called 'Mr Nice Guy totalitarianism' (1993: 159), exploitation and surveillance out in the open, accepted because of the enjoyment of what is received in return. Our enthusiasm for communication, when we allow it to be channelled through thoroughly non-neutral circuits risks becoming enthusiasm for capitalism itself. It may seem peculiar that a system reliant on inclusion and participation is, as indicated above, held to account in this book for spreading social fragmentation and for breaking apart

DOI: 10.1057/9781137394781.0002

bonds of solidarity. But what is desired is simply widespread participation in the circuits of communicative capitalism, the capture of individuals for the agglomeration of information rather than their aggregation as a disparate community or social force. The system moulds workers as communicators, subjectifies consumers as producers, and, on the whole, creates the 'self-facilitating media nodes' once mocked by Charlie Brooker and Chris Morris in the British comedy *Nathan Barley*. This individualisation is a doubling of the atomisation of contemporary work. As Dean argues, here of the USA,

> the experience of consuming media has become progressively more isolated – from large movie theatres, to the family home, to the singular person strolling down the street wearing tiny headphones as she listens to the soundtrack of her life or talks in a seeming dementia into a barely visible mouthpiece. This isolation in turn repeats the growing isolation of many American workers as companies streamline or "flexibilize" their workforce, cutting or outsourcing jobs to freelance and temporary employees. (2009: 4)

This doubling is carried over into surveillance, where the 'Shenzenism' defined by Guy Standing (2013: 133) as the complete surveillance of a workforce, as achieved in the factory-cum-panoptic-town of Shenzen (where the Apple iPhone is built, amongst other things) is repeated by the voluntary mass surveillance of social media users, who tacitly consent to the process by agreeing to terms and conditions it is unlikely many take the time to read.

How do we resist the negative social consequences of a system that elsewhere provides us with so much we enjoy? One of the biggest challenges is that communicative capitalism seems to promote individualism in a way that is incompatible with ideas of solidarity and collective resistance. Social media showcase a society that celebrates the individual, that locates (and corporations extract) value in opinion, commentary and the cult of the lone voice. This atomisation is perversely mirrored by the loneliness of the precarious worker, stripped of collective bargaining power and forced to compete against, rather than co-operate with, other workers for diminishing returns, where worker relations have been marketised by the ascendency of competition over co-ordination in a neoliberal social context – where, in any case, self-facilitating entrepreneurs, not collective labour forces, constitute the ideal and idealised worker. Virilio (2012: 52) argues that 'mass individualism is one of the major psychopolitical questions for humanity in the future'. The purpose

DOI: 10.1057/9781137394781.0002

of this book is to chart a pathological system that incorporates fragmentation in ways that leads to the precariousness of both labour and the social that systematically produces mental fragilities whilst individualising the problem so as to evade responsibility, that capitalises on the social anxieties that it helps to produce through excessive communication, and, ultimately, operates according to a logic of efficiency – in communication, development in general – that is utterly inhuman in its regard for human misery. What follows is intended as a diagnostic text rather than a manifesto, but the overarching concern is with a call for *autonomy over automation*. 'Autonomy', writes Berardi (2009a), 'is the independence of social time from the temporality of capitalism'. According to Michael Hardt, the Autonomists of the 1970s focused on

> the emerging autonomy of the working class with respect to capital, that is, its power to generate and sustain social forms and structures of value independent of capitalist relations of production, and similarly the potential autonomy of forces from the domination of the state. (1996: 2)

Their slogan was *the refusal of work:* not of all productive labour but of productive labour constrained within the relations of capitalist production, autonomy from states, parties and corporations, self-valorisation in the form of a new kind of sociality needed to build a new kind of society (Hardt 1996: 2–3; for a history of the Autonomist worker movement see also Berardi 2007). It was a movement that declared that more work was not beneficial to society, that a massive reduction in work time was needed to free the social from the constraints of capital (Berardi 2009b: 213), to reject poverty shared equally in order to achieve a collective wealth of pleasure (Hardt 1996: 6-7). Technological development ought to have made this possible but instead automation has expanded work whilst diminishing its remuneration; new communication technologies might have invigorated social organisation free from capital but have instead transformed communication, the social, into productive forces. Automation has brought about new regimes of regulation and control that sprawl far beyond sites of work, occupying every facet of our lives by snaking through the circuits of communicative capitalism, achieving the governance of social existence by co-opting the social in work and making work out of the social. This is not to call for the rejection of automation but rather its subordination to social goods and the equal redistribution of the wealth of leisure it opens up, nor for the rejection of technological modes of communication in favour of some Year Zero of co-presence,

DOI: 10.1057/9781137394781.0002

but rather the uncoupling of communication from principles of productive efficiency and the interests of capital. This book, finally, aims to locate some of the key sites for the struggle of *autonomy over automation*. In Chapter 1 it is argued that cognitive labour is inherently precarious and that, with communication now a primary productive force, precariousness is not confined to a given class but is instead a defining feature of social existence under the governance of communicative capitalism. In Chapter 2 it is argued that the acceleration of communication, and the demand to be attentive to hyperactive circulations of information, has created a mentally exhausted, *never not on* workforce, sleep-deprived and labouring under conditions of anxiety and sadness brought about by the precarious conditions of life in a 24/7 global economy. In Chapter 3 it is argued that social anxiety is exacerbated by the organisation of both work and leisure time through circuits of communication that instrumentalise language, circuits of profitable connection but social disconnection that undermine conditions of community and solidarity. And in Chapter 4 it is argued that communicative capitalism extends and accelerates itself according to an ideology of development that is pathological, whilst barriers to and potentials for resistance to this inhuman process are charted. Overall, this book is motivated by a desire to locate the ways in which our communicational systems have become pathological so that we might then begin to question the ways that communication has come to be valued in contemporary society.

DOI: 10.1057/9781137394781.0002

1
Cognitive Labour

Abstract: *This chapter explores the precariousness of contemporary labour, arguing that it is not confined to any given social class but is instead an inherent feature of* communicative capitalism. *Cognitive labour marks a shift in productivity from the body to the soul of the worker, expropriating mental energies as well as putting to work the subjectivity of the worker. It is argued that this is not confined to creative or knowledge work but is instead identifiable in all labour that is communicational, relational or affective. Such productivity is most effectively mobilised by networked technologies in ways that make it flexible, fragmented and insecure.*

Hill, David W. *The Pathology of Communicative Capitalism*. Basingstoke: Palgrave Macmillan, 2015. DOI: 10.1057/9781137394781.0003.

DOI: 10.1057/9781137394781.0003

In his examination of the demonisation of the working class, *Chavs* (2012), the journalist Owen Jones sketches a broad picture of industrial decline in the UK, and with it the substantial replacement of manual work with something altogether less physical. Compared to coal mining or car manufacture, say, these present jobs – retail service, hospitality, care work, customer relations and so on – exert much less of a demand on the body, but they are also less secure, lower status and worse paid (Jones 2012: 145). It is important to consider how the effects of this shift – decreased job stability for many, the uncoupling of work from stable identity and any affective benefits that would previously have accrued, working poverty but also the increasing difficulty of matching work time to remuneration – play out in what has been called communicative capitalism. Most importantly, though, it is necessary to understand the move from physical to mental or cognitive labour since, it will be argued below, this is fundamental to the precarious conditions of work today. To take an example, this from Jones (2012: 147), there were, in the 1940s, almost a million people working down the mines in the UK; today, upwards of a million work in call centres. Jones (2012: 147) identifies in the call centre worker 'as good a symbol of the working class as any' but this can be extended further by identifying the call centre worker as emblematic of the deleterious aspects of work under the conditions of communicative capitalism.

The call centre worker is on the front line of customer relations providing a role that is, therefore, fundamentally social – work is social insofar as it captures forms of cooperation for production, even if there is little sense of dialogue (Virno 1996: 24) – and yet the conditions of their labour, as Jones describes them, are thoroughly uncivil. The job primarily entails communication and yet those doing it are dissuaded from communicating with one another, and little opportunity to do so will exist anyway as each worker is unavailable to the other by the presence of the headset through which they conduct their work. The employees' computers do not simply facilitate their daily tasks but also monitor – inflexibly – their productive output and presence at the terminal, timing the duration of periods away from the screen. In some cases, Jones observes, workers must raise their hands to request permission to take toilet breaks, with only an arbitrary allotment of time allowed for such vital bodily functions. And the nature of communication is ultimately uncreative, as the working day consists of reading to callers (or those cold called) a predetermined script that forecasts their

DOI: 10.1057/9781137394781.0003

possible responses. Overall, the picture received is one of a divestment of autonomy facilitated by constant surveillance. As a workplace, the call centre exemplifies the transition from a Foucauldian space of discipline, to the society of control described by Gilles Deleuze (1992), where the enclosure of the factory is replaced by the modulation of the language of code enacted through the computer. Maurizio Lazzarato (1996) expands on this account of control when setting out what he describes as *immaterial labour*. This form of labour encompasses the informational and computer control – data-processing, knowledge-production, computational work and so on – but also a range of human activities that would not previously have been considered as work at all. As such, activities that now contribute productivity include communication, networking, care, intersubjectivity and social life in general. The objective – or at least the effect – of corporations of control is, then, for 'the worker's soul to become part of the factory' (Lazzarato 1996: 134), as the social qualities of the employee are put to work. Lazzarato argues:

> The management mandate to "become subjects of communication" threatens to be even more totalitarian than the earlier rigid division between mental and manual labor (ideas and execution), because capitalism seeks to involve even the worker's personality and subjectivity within the production of value. (1996: 136)

Our call centre worker, then, is not only subjected to domination exerted over their social existence by the environment of work – the regulation of breaks, the lack of opportunity for communality with fellow workers – but also finds their social nature put to work, a price placed on the appeal of their personality and the efficacy of their communicational ability for the generation of revenue. The scripting of calls constrains personality and communication within a pre-programmed mode, forecasting the conversation with the customer and undermining the employee's autonomy in speaking. As Lazzarato explains:

> The corporation, in certain cases (call centres), diagrammatically exploits even language, by reducing signifying semiotics to a means of signalling that simply triggers prefabricated address and response procedures. There is nothing of the dialogical event in the verbal exchange between employee and consumer. Words and propositions are the "input" and "output" of the machinic enslavement specific to service relations. (2014: 115)

That said, to what extent does such work constitute a form of cognitive labour? Jones (2012: 149) is probably correct in observing that call

DOI: 10.1057/9781137394781.0003

centres do not demand workers to think so much as communicate, and here it might be added that such communication is neutered of its spontaneous social nature. But when Mark Fisher (2011: 6) argues that 'so-called "cognitive labour" has been overstated', that 'just because work involves talking doesn't make it "cognitive"' and that 'the labour of the call centre worker mechanically repeating the same rote phrases is no more "cognitive" than that of someone on a production line', a distinction ought to be made. As Christian Marazzi (2011a: 115) argues, the entry of communication into the realm of production is the exploitation of language, and linguistic ability is the most fundamental form of human mental capacity; in turn, cognitive labour can only be operationalised by communication, since workers cannot remain brains in vats. With cognitive labour it is its immateriality that is of primary importance, the way capital profits not from oil or gas or iron, but from the soul of the worker, that is, the social, human core of their being. If now the call centre worker has replaced the coal miner as a symbolic figure within communicative capitalism, it is their *subjectivity* that is now mined by corporations. The control exerted over the worker is directed at their cognitive faculties, even if one is tempted to say that the form of labour itself may be somewhat mindless.

To maintain that the call centre worker is emblematic of a kind of cognitive labour that takes place in communicative capitalism then what remains to be settled is quite what is and is not cognitive labour. To speak of the *knowledge economy*, or to understand cognitive labour as synonymous with the kind of knowledge work that is understood to take place in such economies, is to focus on one section of the economy at the expense of all others. This is what happens when the idea of a *knowledge economy* is conflated with *Silicon Valley*, losing sight of sectors that are no less immaterial. Similarly, to emphasise the *information society* is to run the risk of confusing information with knowledge, worse, reducing knowledge to information, and therefore of ignoring sectors of work that are reliant on knowledge which cannot be reduced to information (Moulier Boutang 2011: 40). Both draw the attention away from capitalism, specifically the form of communicative capitalism that orders the economic and social realities of labour today. Cognitive labour includes not only information workers – programmers, ethical hackers, financial traders and so on – but also service and hospitality workers, telephone hawkers, carers, and many more jobs besides. Marazzi (2011a: 94) identifies cognitive labour taking place where workers are no longer using machinery that is wholly external (such as factory equipment) but 'technologies that are increasingly mental, symbolic, and communicative', where value exists in the worker herself, 'in her brain, in her soul'. Take the care worker: the valuable role performed here relies on a set of mental technologies organised around re-cognition of the vulnerable other, interpretation in the form of empathy, communication essential to negotiating ailments and also as balm to soothe them. Care work is a prime example of cognitive labour as a role that cannot (yet) be automated, that relies on the reflexive judgement of the human mind. The decline of industry has not led to a decline in live labour but rather, after mechanisation, a kind of globalised 'Mexicanization' (Marazzi 2011b: 91), that is, the proliferation of low-skilled, low-paid mental labour. Cognitive labour is a sliding scale of mental work, peaking with the technologists and the creatives and the traders of information, but bottoming out with forms of work whose basic tool is communication and which cannot yet be performed without the specific benefits of human intellect in its most general and unremarkable form. Language, emotion, empathy, recognition, communication: these are the required skills for labour in communicative capitalism. We are witnessing what Bernard Stiegler (2011: 21) calls 'the proletarianization of the human mind', or, more accurately, the extraction of value from the human nervous system. *Mexicanization* is the failure of autonomy over automation.

This process might be understood as creating what Franco Berardi (2009a: 8) calls a 'cognitariat', or a mass of mental labour that exists not as a collective – as a body of cognitive labour that can be organised – but as a mass of intellectual productivity that is created by individual and largely isolated workers. Another way of thinking of this kind of mass is as an archipelago of un-automated, non-autonomous labour, a cluster that can be utilised as a coherent whole but wherein each labourer or act of labour is an island to itself. Whilst all labour requires – and always has required – intelligence and is cognitive in nature, Berardi (2009a: 34) argues that today 'cognitive capacity is becoming the essential productive resource'. It is no longer possible to see communication as something that merely happens alongside productive labour, as when, say, workers might steal snatches of conversation amongst themselves on the production line. Virno (1996: 16) goes so far as to say that chatter, once something seen by employers as idling and appropriate only at the end of the working day, is now regarded as being productive itself in the workplace – he is thinking here of the modern office – where gains can be made by a sort of

DOI: 10.1057/9781137394781.0003

chattering opportunism rather than mute instrumentality, problem solving and idea generation through a no-holds-barred group conversation. In the factory, amidst the loud machinations of heavy plant, communication was unproductive insofar as it would necessitate the shutting-down of the tools of labour so that workers could hear one another; now, it is not sustainable to oppose communicational – and relational – work to productive labour. Marazzi (2011a: 54–55) illustrates this point well with his example of UPS, the parcel delivery service in the United States. In an attempt at increasing the efficiency of deliveries, the number of drop-offs each driver could make per day, UPS decreased the amount of time each driver was allowed to take getting the parcel to the customer. This, effectively, was a rationalisation of the face-to-face, a decrease in the amount of driver-customer interaction at the doorstep. What UPS discovered was that less face-to-face time turns out to be inefficient; customers had valued the conversations with drivers, buying into a kind of affective relationship with the company through its employees, but also gleaning information about new services that they might use. In the end, UPS increased the amount of time given over to communication between drivers and customers, the lesson learned that talk may be cheap – in the sense that workers are not necessarily financially remunerated for this – but it is also productive. Unlike at the call centre, this communicative labour does not come uppermost on the job description, but it demonstrates how it is nonetheless significant in terms of profitability. Any kind of work that is relational mobilises language such that, most often, work is indistinguishable from speaking (Lazzarato 2014: 113).

That said, it is the development of increasingly effective information and communication technologies that takes communication from being a part of labour to being the primary productive resource. What Jodi Dean has called communicative capitalism, Berardi (2009a: 18) instead calls 'semiocapitalism', understood as a 'fusion of media and capital', where the communicational and relational capacity of the worker is made productive through cognitive labour. By focusing on the exchange and value of signs Berardi's attention is primarily drawn to the dominance of forms of productive communication that are enunciated beyond the everyday language of sociability, an account that will be explored in Chapter 3 insofar as this less human form of communication runs against collectivity, solidarity and social empathy, and that is, here, useful for thinking about how immaterial forms of labour are abstracted from the body of the worker through technology. Berardi (2012: 19) writes: 'The word is no

DOI: 10.1057/9781137394781.0003

longer a factor in the conjugation of talking bodies, but a connecter of signifying functions transcodified by the economy'. If cognitive labour is largely the sorts of tasks that are left over after automation, then through communications technologies is achieved an abstraction of communication from its intersubjective function, as those tasks increasingly involve communicating information over exchanging meaning. Ironically, the tasks left to humans by computers are carried out in such a way that we might speak of the automation of thought (Berardi 2012: 28); computers may not be able to do what we can do, but the way we do it is constrained into modes more readily recognisable as machinic than as human. In place of 'the conjugation of talking bodies', interrelation between workers, there is an abstracted co-operation of networked human brains – *thought without a body*, if only in the sense that the body gives way to an archipelago of cognitive energy. As Marazzi (2011b: 57) observes, this abstraction is vital for creating value 'beyond the separation of company and territory, between public and private spheres, between individuals and organization'. Networked communication technologies dematerialise not only labour, but also the workplace (insofar as work no longer requires a dedicated space), the divide between working time and leisure time, between the company and its exterior.

This chapter has so far attempted not to ignore forms of work that do not meet some imagined expectation of creativity or minimum level of intellectual ability, focusing instead on communication as essential to productive labour. The argument here is that, in so far as the primary resource in this socio-economic environment is the *general intellect* of populations, the temptation to identify cognitive labour with any one, or narrow range of, sector/s of work must be resisted. However, in understanding the primacy of communication it is essential to look to work that is networked through communication technologies because the mass mobilisation of cognitive labour is unimaginable without the flexibility of work that this dematerialisation makes possible (Marazzi 2008: 150). The general intellect describes the collective knowledge and intelligence of a population or society in a given moment of time (see Virno & Hardt 1996: 262). The concept is adapted from Karl Marx's Fragment on Machines in the *Grundrisse* (1993), although no longer embodying his exact meaning, by thinkers such as Marazzi, Lazzarato and Berardi, put to work to describe the way that corporations today exploit distributed networks of labour, bulked out with reserve labour, in order to maximise productive efficiency, innovation and profitability. As Marazzi (2008: 93)

DOI: 10.1057/9781137394781.0003

defines it, the 'entrepreneurial conjugation of the *general intellect* consists in transforming communication into an assembly line, turning speed and productive and distributive interconnection into commodities'. Yann Moulier Boutang (2011: 34) suggests more concisely that it consists of 'creativity distributed through the entirety of the population', understanding the general intellect as the primary resource for contemporary capitalism. The general intellect is a loose network of workers, increasingly accessible to corporations by networking technology. The employment of the general intellect is modular, the network allowing the corporation to access the productive power – communication, creativity, social network and so on – of vast swathes of a given population, but not all at once. The way that the corporation exploits the general intellect is *just in time*, utilising sections of the network only as and when they become useful. As Berardi (2011: 130) puts it, somewhat more colourfully, the general intellect put to work has constituted 'an infinite brain-sprawl, an ever-changing mosaic of fractal cells of available nervous energy'. As such, the general intellect, this agglomeration of cognitive labour, affects the stability and duration of labour itself, a point that will be addressed presently. A prime example of the general intellect is provided by Amazon's Mechanical Turk, a website that operates as a market place for discrete units of networked labour. The website is marketed as providing 'businesses and developers access to an on-demand, scalable workforce' that is 'global' and available '24 × 7' whilst workers 'select from thousands of tasks and work whenever it's convenient' (Amazon 2014). What is being offered, then, is a network for units of fragmented, just in time labour. The labour being sought and offered is defined as 'Human Intelligence Tasks', that is, work that cannot currently be performed by computers. Examples include locating information for inclusion on websites, extracting summaries of information from datasets, recording text displayed on images and so on. This is work that requires human intelligence over machine processing – cognitive labour – without necessarily requiring a great deal of that intelligence. The majority of cognitive labour is precisely work that requires human intelligence in this basic sense, the remainder left by the machines, put to work largely through networks as corporations seek to mine the general intellect.

If cognitive labour can be identified across the modes, rather than concentrated in the most desirable or well rewarded forms of employment, then what remains to be argued is that by virtue of the flexibilisation of work that goes hand in hand with it, precariousness is not

DOI: 10.1057/9781137394781.0003

concentrated at the bottom of the social heap, but endemic to communicative capitalism. In 2013, the BBC released the results of its extensive Great British Class Survey. Undertaken by a team of sociologists (Savage *et al* 2013), the survey considered economic, social and cultural factors to identify a new class system in Britain comprising seven groups:

1 The Elite
2 The Established Middle Class
3 The Technical Middle Class
4 New Affluent Workers
5 The Traditional Working Class
6 Emergent Service Workers
7 The Precariat

Whilst space does not allow for a full analysis of this class picture here, the identification of precariousness as inherent to a bottom-rung class merits attention. The journal article that explains the methodology and results of the survey describes the precariat as 'economically the poorest class', with 'negligible savings', most likely renters with a small social range, located in former industrial areas, and 'the most deprived of the classes' identified by the study (Savage *et al* 2013: 25). They go on: 'Occupationally they are over-represented amongst the unemployed, van drivers, cleaners, carpenters, care workers, cashiers and postal workers, and they also include shop-keepers' (Savage *et al* 2013: 25). Whilst the class described is undoubtedly precarious to a high degree, by identifying precariousness so closely with one group the impression is given that it is bound up in socio-cultural-economic division rather than being inherent to the kind of labour that is primary today. It also goes against popular conceptions of the nature of precariousness. For example, Marazzi (2011b: 30) identifies precariousness as a result of de-industrialisation and the resultant 'reduction in the cost of labour, attacks on unions, automisation of entire labour processes, delocalisation to countries with low wages' and so on. The Great British Class Survey speaks to one half of this story by identifying the precarious lives of those in former industrial areas, but not the wide-scale transformation of labour by the capture of mental, relational and communicational energies. Michael Hardt highlights the role of globalisation, by which

> capital is moving away from dependency on large-scale industries towards new forms of production that involve more immaterial and cybernetic forms of labor, flexible and precarious networks of employment, and commodities increasingly defined in terms of culture and media. (1996: 4)

DOI: 10.1057/9781137394781.0003

Federico Campagna locates precariousness within a general malaise at the failure of governments across Europe to foresee, prevent or cope with the financial catastrophe of 2008, making it at once social, economic and political:

> the whole of Europe is shaken by the closure of traditional lines of financial credit and the simultaneous, forced extension of the line of credit on the lives of workers, in the context of a generalized indifference of the State for the well-being and consensus of its own citizens. At the same time, the collective body and brain of the people of Europe is traversed by a growing fear of the precarity of the material existence, a melancholia for a lacking 'order', a widespread and abstract feeling of injustice, a desire for a quick and harsh punishment of whoever is to be deemed 'responsible' for the current state of things and, on top of this, a not too subtle desire for a retreat within the 'imagined communities' of modern Nation States (2011).

Tracey Jensen (2012) draws our attention to the narrative of post-crash austerity politics, which

> ignores the intensified precarity of *all* labour – the rise of short-term contracts or contractless work, underemployment, low wages, the threat of outsourcing, diminishing returns on maternity pay and sickness pay, the failure to recognise caring responsibilities, 'flexploitation', the shift of education and training costs to the individual and so on.

By glossing over these factors – which it would be difficult to define as threats to just the lowliest of classes – austerity measures can get to work in dismantling the welfare system to reduce the national deficit, at the same time furthering precariousness by removing the safety net if one should fall (ill, pregnant, unemployed and so on). Whilst Campagna and Jensen make the link between an all-pervasive precariousness and the financial crisis, Berardi (2012: 8) goes so far as to argue that the entire financial system – pre- and post-crash – is 'based on the exploitation of precarious, cognitive labor: the general intellect in its present separation from the body'.

Savage *et al* (2013: 25) note that the use of the term precariat is faithful to that of Guy Standing in his book *The Precariat: The New Dangerous Class* (2013). Here Standing (2013: vii) identifies his own class structure, with the precariat once more at the bottom of the ladder:

1 The Elite
2 The Salariat

DOI: 10.1057/9781137394781.0003

3 Proficians (professional technicians)
4 The Working Class
5 The Precariat

Again, space does not permit analysis of the complete system but it is sensible to turn to the source to ask whether it is useful to map precariousness onto a given class. Standing (2013: 1) notes that students, migrants and those with unconventional lifestyles are the most visible members of the precariat but argues that there are many more besides. The idea that the precariat is not a homogeneous group but that all those in it 'share a sense of their labour as instrumental (to live), opportunistic (taking what comes) and precarious (insecure)' (Standing 2013: 14) is convincing. However, it is how he attempts to box off this group that does not seem to add up:

> The precariat was not part of the 'working class' or the 'proletariat'. The latter terms suggest a society consisting mostly of workers in long-term, stable, fixed-hour jobs with established routes of advancement, subject to unionisation and collective agreements, with job titles their fathers and mothers would have understood, facing local employers whose names and features they were familiar with. (2013: 6)

Since Standing retains the working class in his structure, the idea that it is characterised by stability is as questionable as the idea that insecurity is not inherent beyond this. Indeed, this ascription of stability seems both arbitrary – there is no argument put forward to support it – and unsustainable with reference to Richard Sennett's (1999: 76–97) account of risk and employment, an account that predates our most current financial crisis by a decade, in *The Corrosion of Character*. Sennett describes conditions of employment where every job is started anew day on day, since no one remembers previous accomplishments or, in any case, they are given no weight; where moving jobs can land employees worse off than before, even when it looks like an upwards move (but is often sideways); where training and education is a gamble that can leave people over-qualified for the remaining jobs once they have been out-competed at the higher end; and where growing old is precarious in itself, since age is associated with rigidity, youth with flexibility, and it is easier to employ the latest batch of technologically literate youngsters to enter the job market than it is to retrain existing, older employees. The subjects of his report were not carpenters or postal workers but advertising copy writers and IT consultants.

DOI: 10.1057/9781137394781.0003

Even before Sennett's study, Marazzi had identified the financial crisis of the early 1990s as a white collar recession (2011a: 122), where a post-industrial middle class, who had attempted to capitalise on its resources of knowledge and creativity was, in so doing, left open to the vicissitudes of the market (135). Being self-employed, or engaged in small entrepreneurship, or office temping, for example, brings volatile precariousnesss. Rosalind Gill and Andy Pratt (2008) have also demonstrated the precariousness of creative and cultural work – forms of immaterial labour highly regarded in terms of their profitability and as motors of economic regeneration for policy makers – where temping, internships, freelancing, short-term contracts, high churn, non-traditional pay and so on, erode any sense of stability. Whilst not all cognitive labour is creative, it is also the case that precariousness is not limited to the van drivers.

At a more basic level, the ascription of precariousness as an identifier for the poorest in society does not seem to tally with the spirit of the word itself. Standing (2013: 24) argues that the precariat is growing and could potentially swallow up any of us from outside the elite, writing: 'Many of us fear falling into the precariat or fear that our family and friends will do so'. Later he cements the point: 'Falling into the precariat could happen to most of us, if accidents occurred or a shock wiped out the trappings of security many have come to rely on' (2013: 59). Precariousness, however, is precisely the state of insecurity or lack of fixity that makes falling a live and dangerous possibility. 'Precarious' is a causal power or disposition, the potential for one ontological state of affairs to become another. If many of us have not only a fear of falling but also a realisable chance of falling then it seems prudent to say that many of us have precarious existences. In arguing against the idea that precariousness unites a class, Angela Mitropoulos (2006) suggests that to be precarious is to teeter – something that Standing acknowledges but cannot seem to tally with his thesis that the precariat is a dangerous new class (see also Seymour 2012; Woodcock 2014). If the so-called precariat is the most destitute group in society then the issue is not so much precariousness as it is a social catastrophe, since they have ceased to teeter and have instead fallen into poverty. The most important point to emphasise is not a class concentration of precariousness but the way that, with the present conditions of employment, precariousness is common to many and inherent to forms of labour in communicative capitalism.

DOI: 10.1057/9781137394781.0003

Standing (2013: 1) acknowledges that neoliberal market flexibility has created a global precariat 'of many millions' but since flexibility is essential to the exploitation of cognitive labour across the board then this seems a gross underestimation. The opposite of the flexible work that is so heralded today is, in fact, worker autonomy. Berardi (2009a: 75) defines autonomy as the freedom of social time from the deleterious rhythms of capitalism. Thinking back to workers' struggles in the 1970s, primarily in his native Italy, Berardi recalls the withdrawal of labour, sabotage and personal hardship faced by workers seeking an escape from the sadness of work, and draws a forlorn conclusion: 'Workers demanded freedom from the lifetime prison of the industrial factory. Deregulation responded with the flexibilization and fractalization of labor' (2009a: 76). The workers had looked, in particular, to technology to free them from labour; instead it freed capital by making production immaterial. The result was downsizing, outsourcing, short-termism, a general scramble for hours and greater insecurity. Flexibility is not freedom for the worker; flexibility is precariousness and, ultimately, a prison of fear. As Berardi (2012: 75) argues: 'Precarization is not only the loss of a regular job and a salary, but it is also the effect of fragmentation and pulverization of work'. The modern worker is expected to be mobile, to retrain and adapt constantly, to command flows of information, and so on (Virno 1996: 14). It is the worker who has to be flexible, more so than the work itself. Job security is largely a thing of the past. The old way of thinking of unemployment as the obverse of employment does not hold any more; increasingly, our lives consist of short periods of employment, unemployment and training (Lazzarato 2014: 153): more fluid and, as a result, more precarious. It is a long time since Sennett (1999) identified the then prevailing business mantra of *No Long Term*; if then it seemed like a war cry to an assault on rigidity, today the battle is won and it is scarcely imaginable that it could be anything other than descriptive. Even then, Sennett (1999: 9) could see the writing on the wall, that the idea of a *career*, meaning a fixed road, no longer made much sense, replaced by the less fulfilling *job*, meaning a lump of something, since that is what so many now do, lumps of labour, often unconnected. Such a situation is economically perilous for the individual, but it also leads to a less stable sense of identity, since there is no long term career to offer a referent for public self-definition (Bauman 2004: 27). Flexibility itself is precarious.

DOI: 10.1057/9781137394781.0003

With increased flexibility we get, not worker autonomy, but increased worker responsibility. This responsibility is in fact an outsourced mechanism of control:

> Capital wants a situation where command resides within the subject him- or herself, and within the communicative process. The worker is to be responsible for his or her own control and motivation within the work group without a foreman needing to intervene, and the foreman's role is redefined into that of a facilitator. (Lazzarato 1996: 136)

The immaterialisation of work means the foreman has disappeared, internalised by the worker; it also makes the contours and parameters of work and the company impermanent and less certain. As Sennett (1999: 19) observed, this responsibility is a burden of taking control when one has little authority. Today, many workers are left to organise their own working time, either the hours they choose or the tasks performed within that time, lacking the security of a routine or a structure. The more control workers are forced to assume, the greater the risk of losing control, of messing up – of being fired. Without the rigid bureaucratic structuring of work time it is little wonder that work now seeps out inexorably into social life. We can no longer confidently maintain a distinction between work time and leisure time, since there is always the potential for the latter to be colonised by the former. In his famous essay on idleness, Bertrand Russell (2004) demonstrates the forlorn hope that technological development will lead to reduced work time by supposing the invention of a machine that could double the amount of pins that workers could produce in an eight hour day. Since pins can hardly get any cheaper there is no reason to believe that a reduction in price would shift greater numbers. As such, the sensible course of action would be to reduce the working day to four hours for the entire workforce without, since levels of productivity are maintained, reducing pay. Of course, this does not happen: the eight hour day remains and half the workforce is laid-off. Everyone could have had four more hours of leisure time in the working day but instead half have no more and half have the miserable leisure time of unemployment because, as Russell argues, work is rewarded not in proportion to what is produced but in proportion to the industry exerted by the worker. And, at the root of this for Russell, is the distaste with which working class leisure is received by the ruling classes. However, the situation

DOI: 10.1057/9781137394781.0003

today is different: on the one hand, leisure time for workers is desirable as, in a consumer society, it has become profitable, understood as a kind of work outside of employment (Stiegler 2011: 22); on the other hand, pay is tied neither to the productivity nor to the industry of the worker since (a) cognitive labour in a networked society is difficult to measure and (b) excessive industriousness is not necessarily rewarded, as with unpaid overtime, out of hours emails, lunching at the desk, and so on. The issue of lunch breaks may seem trivial compared to exploitation through unpaid overtime but they are both indicative of the same problem. Stiegler (2011: 53) reminds us that the word *negotiation* has at its Latin root *otium*, or studious leisure. This is not to be understood as frivolous idling but as an essential time-space of care for the self. The *negotium* is, then, the negation of studious leisure and care for the self, when one's existence is negotiated with an employer. Russell (2004) argues that the butcher and the baker are praised for making profits but when the worker enjoys the product of this labour at lunch then this is regarded as idling. Eating is a waste of time regarded as useful only insofar as it is fuelling for productive labour. As such, it is remarkable that, as reported by Jones (2012: 145), many supermarkets in the UK do not pay their workers over their lunch break. Time for the worker to sustain their productivity through maintaining vital bodily functions is forced from the *negotium* into the *otium*, from part of the workday into leisure time. The lunch non-break is the thin end of the wedge of work reconfigured as 'duration without breaks' (Crary 2013: 8) in globalised, communicative capitalism. And has any technological development been more detrimental to the maintenance of a distinction between work time and leisure time than the smart phone? Berardi makes the point eloquently:

> The mobile phone makes possible the connection between the needs of semio-capital and the mobilization of the living labor of cyber-space. The ringtone of the mobile phone calls the workers to reconnect their abstract time to the reticular flows. (2009b: 193)

The Internet has extended work beyond any location in particular, but it is carrying around access to it – in our pockets, to our lunch dates and our so-called after-work drinks, back to our sofas and up to our bedside tables – that has allowed work to take place at any time as well as at any location. Workers today are always networked, always accessible and

DOI: 10.1057/9781137394781.0003

so always on. Our devices have off-switches but what self-responsible, internalised-foreman would dare to use it? Cognitive labour, most effectively mined by networked communications technologies, is marked by a *never not on* demand for work. Our leisure time has become vulnerable – precarious, even.

The problem of work and pay under conditions of communicative capitalism is not only a question of time but also of the immateriality of cognitive labour. Whilst the productivity of such work can be measured in terms of outputs, the work itself is abstract. Marazzi (2008: 43) argues that 'the putting to work of the cognitive properties of the workforce' – language, relational networks, Human Intelligence – 'leads to the *crisis of measurability* of single work operations'. What is a unit of thought? Of care? Of social commerce? When does such work begin and end? Communication technologies mean we are always connected and available to work but it is also the case that mental tasks do not simply end in a clear-cut fashion; it is not as easy as just stopping thinking about something, since rarely do thought processes operate in such a discrete fashion. With cognitive labour, payment for work becomes artificial (Moulier Boutang 2011: 133). It might be attractive to think that what is being paid for is presence at a work terminal, a home computer, a call centre headset and so on, but the productive labour is of the mind, expressed through communication. A contract that states that a worker must perform 35 hours of mental labour a week is meaningless, since not all cognitive labour hours are equally productive or even comparable, and since we cannot separate out mental work from mental life itself. There is no reliable mechanism for placing a value on cognitive labour, such that remuneration for work is arbitrary and volatile.

Ultimately, all forms of labour are inherently precarious and always have been; the only things that can ensure security are external. It is possible to identify a sea change brought with de-industrialisation and flexibility, but the previous security was only held in place by the political force wielded by the collectivisation of workers, allowing for the establishment of rights and employment laws. As Berardi (2009a: 31–32) puts it, the violence of capital was only ever held at bay by the threat of worker violence; with the decline of this political force 'the natural precariousness of labor relations in capitalism, and its brutality, have re-emerged'. What is new is how cognitive labour now constitutes this precariousness. Precarious work – flexible, intermittent, short-term, zero-hours,

DOI: 10.1057/9781137394781.0003

and so on – allows for better utilisation of the general intellect because it expands the network through more workers (Moulier Boutang 2011: 132–33). At the same time that cognitive labour becomes primary within the economy, precariousness becomes widespread throughout society, as the flexibility that immateriality allows leaves the majority prone to uncertainty.

DOI: 10.1057/9781137394781.0003

2

Communicative Disease

Abstract: *This chapter explores the overproduction of communication and information in* communicative capitalism *as well as the need to focus on the structural violence of the mental health crisis created by this system. It is argued that cognitive labour involves an excessive demand on workers' attention, leading to attentive stress, whilst the 24/7 nature of the global economy leads to sleeplessness and compliance. The precariousness of this kind of work erodes self-esteem and provokes anxiety, whilst both governments and pharmaceutical companies are on hand to medicate a depressed workforce in order to maintain productivity despite the human costs.*

Hill, David W. *The Pathology of Communicative Capitalism.* Basingstoke: Palgrave Macmillan, 2015. DOI: 10.1057/9781137394781.0004.

In the final episode of the second series of *Breaking Bad* we see John de Lancie's character, Donald Margolis, return to work after the death of his daughter, Jane, from a heroin overdose. *Breaking Bad* is the story of a chemistry teacher, Walter White, who turns to crystal meth production in order to pay the medical bills for his cancer treatment or, in the event that he fails to make a recovery, to leave some money for his family. Whilst much can be made of the precarious state of medical provision in the United States, where treatment can be prohibitively expensive for those without insurance or for those whose cover is insufficient, where illness can mean a choice between poverty or death, the scene where a grief-stricken, sleep-deprived Margolis returns to his job, mentally and physically exhausted, speaks eloquently to the mental precariousness of cognitive labour. We watch as he sits at a screen, interpreting data and communicating it via his headset. He reads out a stream of barely intelligible technical language, numbers that seem divorced from any kind of experienced reality, whilst watching abstract shapes flicker and move across his screen. That is, until Margolis, our barely functioning air traffic controller, crashes two passenger jets into each other over residential Albuquerque, New Mexico.

In this example, the mental strains of cognitive labour in communicative capitalism exasperate an already existing disposition, the mental health impact of grief. This chapter examines the kinds of strains that work today places on the worker by virtue of the shift of emphasis from the physical to the cognitive and communicative. For example, Owen Jones (2012) reports that the Royal College of Speech and Language Therapists have raised concerns about the increasing number of call centre workers that require referral to speech therapists after losing their voices. Long hours spent communicating with little opportunity to even take a drink of water has exhausted the worker's capacity for productive labour. On the one hand, there is nothing new here; a construction worker who damages his or her back lifting heavy loads is no less incapable of work *because of work* than the voiceless call operative. On the other hand, we might want to say that the loss of speech – or, as will be seen, mental impairment – is, whilst no better or worse than physical injury, significantly different in so far as what communicative capitalism exploits is cognitive capacity, what it therefore controls is subjectivity and so what it damages when it exhausts this valuable natural resource is fundamental to what it means to be human. Take, for now, the lost voice of the call centre worker; what is exploited are the personalities and

DOI: 10.1057/9781137394781.0004

social skills of the employees, their personability put to work through communication technologies; what is exhausted, albeit temporarily, is their ability to communicate through spoken language, this fundamental dimension to our nature as social beings. Christian Marazzi would see this loss of voice as part of the consumption of mental and social resources during cognitive labour:

> In the post-Fordist context, in which language has become in every respect an instrument of the production of commodities and, therefore, the *material* condition of our very lives, the loss of the ability to speak, of the 'language capacity', means the loss of belonging in the world as such, the loss of what 'communifies' the many who constitute the community. (2008: 131)

Here we see how 'the *private* use of the *general intellect* clashes with its *social* nature' (Marazzi 2008: 132), exhausting workers' social capacity, their being in the world and being with others, by destroying in production the communicative faculties that are put to work to make a profit. The general intellect may be in good health, insofar as it is available to serve capital, but the individuals that constitute it are increasingly exposed to communicative disease: stress and exhaustion as a result of the demands of communication and information processing; hyperactive stimulation and excessive strains on attention; sleeplessness brought about by an insomniac 24/7, global market; and the depression and anxiety that comes with the precariousness of cognitive labour.

The general intellect is a constantly productive resource for capital to exploit. By using networked technologies to expand the potential workforce throughout entire societies and across borders, labour need no longer be confined to a specific group on retainer, or a specific geography of work. The general intellect as such has the potential to be infinitely productive, but in practice its growth is limited by the fragile embodiment of labour, 'limits of attention, of psychic energy, of sensibility' (Berardi 2012: 77). At the level of the individual worker, there are limits to how much cognitive or emotional energy can be exerted in work, and when capital trespasses across such limit-horizons it exhausts this resource, and heaps stress upon those it exploits. In the previous chapter it was argued that precariousness might be understood as encompassing a lack of autonomy over work time, and when workers cannot switch off from labour their mental faculties are submitted to a 'competitive pressure' and 'a constant attentive stress' (Berardi 2009a: 34). This exhaustion is no longer purely physical, as with industrial labour, but an exhaustion of the

DOI: 10.1057/9781137394781.0004

brain; even the factory owners so negatively characterised by Bertrand Russell (2004) saw the need for their employees to have leisure time – just so long as it was restful, and set the worker up for more physical exertion in the next working day; but it is as if a shift to cognitive labour, with its obliteration of the work-life balance, has obscured the idea that thinking, communicating and various affective capacities, need time for replenishment. Without it, something will break; work-related stress is the embodied limit-horizon pushed to a dangerous extent – and once the point of its fragile elasticity has been passed, the danger is that stress gives way to psychopathology. Capitalism has always been based on the exploitation of psychic energy, and labour has always been to varying degrees cognitive; but under the conditions of communicative capitalism, where cognitive labour is primary and mined through the general intellect, it 'has subjugated the nervous energy of society to the point of collapse' (Berardi 2012: 66). This risk of collapse is precipitated by the near constant communication that is endemic to the work environment. Take, for example, office emails, a non-stop accumulation of things to respond to, and the attendant decisions that need to be made about which to respond to and in what detail, who to include in the response, and so on – it can be exhausting. And then, when a response is made, the demands and stress of communication is transmitted to co-workers (Dean 2012: 144). The email carbon copy (CC) quickly escalates communicational stimuli, communicating stress at the same time, circulating exhaustion. 'The communicational circuits of contemporary capitalism are loops of drive', writes Jodi Dean (2012: 144), 'impelling us forward and back through excitation and exhaustion'.

Immaterial labour is inherently excessive. In Chapter 1 it was noted that immateriality leads to a disjuncture between labour and its measurement. Without being able to adequately measure production – or its remuneration – there is simply too much work being done, over too many hours. Where labour is primarily cognitive, and therefore most notably communicative, and in particular where it takes place through networked communication technologies, we see what Berardi (2012: 150) calls 'semiotic inflation', the excessive acceleration of the production of signs. Within work cultures that encourage a *never not on* disposition towards labour, workers are placed under demands for constant attention to flows of information and lines of communication, a subjection to intensified stimuli that is thoroughly over-stimulating. Even something as innocuous as computer loading times contribute to this: when they

DOI: 10.1057/9781137394781.0004

were slow, workers had more time to relax; now that they are rapid – and increasingly so – there is an increased density of work time (Moulier Boutang 2011: 74). Tiziana Terranova (2004: 1) writes that network culture is characterised by 'an unprecedented *abundance* of informational output and by an *acceleration* of informational dynamics'; where this culture is adopted in the workplace, the result is an all-encompassing excitation of mental faculties. This overload might best be understood as a coming crisis of psychic over-production:

> The mental environment is saturated by signs that create a sort of continuous excitation, a permanent electrocution, which leads the individual, as well as the collective mind, to a state of collapse. (Berardi: 2011: 94)

Paul Virilio (1998: 89–102) classifies informational over-stimulation as a form of eye pollution but this does not quite get to the heart of the problem. Certainly, there is merit to his argument that we are witnessing an 'industrialization of vision' (1998: 89), over-exposure to signs and images as part of the industrial production of immaterial commodities. He warns of a coming 'pathology of perception' (1998: 90), the risk that this over-stimulation will render us (metaphorically) blind, but the real pathology to confront is one of attention. The perceptive field may be polluted, but the important thing about pollution is its deleterious impact, the natural resources it depletes, and it is more importantly our ability to pay attention, rather than that of vision, that is being used up. 'If information is bountiful', observes Tiziana Terranova (2012), 'attention is scarce because it indicates the limits inherent to the neurophysiology of perception and the social limitations to time available for consumption'. The precariousness of the brain, rather than any one sensorial function, is what is brought about by this information overload. Human attention is a scarce commodity; it also has diminishing returns, that is, it is a perishable commodity, and information and communication consumes attention such that a wealth of information leads to a poverty of attention (Marazzi 2008: 66). As Marazzi (2008: 67–68) observes, if workers increase their attention then their work time increases with it (rather than decreasing), and if workers do not increase their attention then their income decreases. As such, there is a competitive demand to essentially exhaust one's reserves of attention. The speed of information is wrecking our attention spans; if attention is a scarce resource, then the acceleration of technological development means that it will be used up at the level of the individual

DOI: 10.1057/9781137394781.0004

worker at an ever greater rate of speed – until ultimately exhausting the general intellect. The danger is that when forms of labour are in some way immaterial, or involve production of commodities that are to some extent immaterial, we lose sight of the physical resource behind them – the human brain. We must not forget that 'cognitive workers, in their concrete existence, are bodies whose nerves become tense with constant attention' (Berardi 2009b: 105), a kind of exhaustion that encompasses the worker's entire being, rather than just their bodies. The brain is unlike the other muscles of the body, insofar as it is constantly working. If productivity is to be increased then this means an increase in stimulation of the human brain – and the danger is that this will cause psychic damage. A number of theorists have begun to speculate that the attentive stress of communicative capitalism, the way that cognitive labour leads to an attention deficit for workers, somehow maps on to the identification and medicalisation of Attention Deficit Hyperactivity Disorder (ADHD) (see, for instance, Marazzi 2008; Crary 2013; Fisher 2009; Virilio 2012). Jonathan Crary (2013: 56) points to a correlation between the intensification of performance and competitiveness in the workplace and the need for ADHD treatment drugs, such as Ritalin, to be prescribed to workers. Mark Fisher (2009: 25) goes further, arguing that ADHD is a result of communicative capitalism, of being wired into hyperactive communication flows. In a similar vein, Virilio (2012: 92) redefines ADHD as a disorder of *hyperinteractivty*, a result of the reconstitution of the social world and the world of work as always on, in part something hitherto unimaginable without networked, interactive technologies. He finds it perverse that children with ADHD face social stigmatization despite being 'in unison with the mad rhythm' of life under conditions of communicative capitalism (2012: 92). It would be reckless to attempt to ground a causal relation between conditions of work and ADHD, and these authors appear to be working at the level of metaphor rather than that of bad science. The question that is important in this respect is not *Does communicative capitalism cause ADHD* (how would one set about investigating this, let alone proving it?) but, rather, *Why do we as a society agree that ADHD is a psychopathology and yet do not extend this thinking to conditions of work that use up resources of attention and, through informational and communication over-stimulation, are inherently hyperactive (or hyperinteractive)?* In this respect, cognitive labour bears all the hallmarks of a pathological disorder and the mental pollution of communicative capitalism is akin

DOI: 10.1057/9781137394781.0004

to brain damage. The danger is that we might all be too distracted, and too mentally drained, to do anything about it.

If only we could sleep on it, recharge our batteries and recover that mental energy. But this constant attentive demand creates an exhaustion rollover, as diminished time for sleep does not allow for adequate recovery from mental exhaustion. Guy Standing argues that

> the global economy has no respect for human physiology. The global market is a 24/7 machine, it never sleeps or relaxes; it has no respect for your daylight and darkness, your night and day. Traditions of time are nuisances, rigidities, barriers to trading and to the totem of the age, competitiveness, and contrary to the dictate of flexibility. If a country, firm or individual does not adapt to the 24/7 time culture, there will be a price to pay. (2013: 115–16)

Efficient exploitation of the general intellect through flexible, cognitive labour creates a precarious time culture that is at odds with restfulness, which employs a duration of labour that is not adequate for restorative sleep. The sleeplessness of work in a 24/7 economy, coupled with the informational and communicational over-stimulation that encourages an emotional over-investment in work, and that distracts from the leisure time that ought to bookend and sustain the working day, but that is under-siege and now too freely yielded, constitutes workers as insomniac subjects. Crary (2013: 10) describes sleep as 'an uncompromising interruption of the theft of time from us by capitalism', which is to say that no value can be extracted from sleep, it is, as such, fundamentally unproductive. Cognitive forms of labour seek to exploit thinking, communication, feeling and so on, and yet the very thing that sustains the effectiveness of these modes of labour is seen as a barrier to work rather than its guarantee, and so less time is afforded to it. Sleep deprivation leads to helplessness and compliance – which is perfect for extracting information, as evidenced by its use in interrogation and torture – but also, ultimately, to psychosis, neurological damage and death (Crary 2013: 6–7). The sleeplessness of communicative capitalism is pathological. But it is also useful if compliance in the work place is desired, since an exhausted worker will do what they are told. A compliant workforce labours without break, makes themselves always available as minds for the extraction of information, of value, so perpetuating a vicious circle of exhaustion whose end only arrives through a crisis of sleep that takes the form of a neurological crash, a stimulation boom that ends in a psychic bust.

DOI: 10.1057/9781137394781.0004

Worker self-esteem is a hindrance to employers who seek not only compliance but obeisance to their demands, yet, like attention and sleep, this too is a scarce and perishable resource. Decline in the status of jobs available to precarious labourers undermines workers' sense of self-worth, and so too the quality of their lives (Jones 2012: 160). Performing Human Intelligence Tasks that no one has yet worked out how to automate is hardly fertile ground on which to develop confidence. Workers can have little understanding of their own social worth if self-esteem cannot be derived from labour. This is somewhat ironic given that cognitive labour is primarily geared towards extracting value from labour's social being. But at the same time that social skills become profitable – communication, affect, networking, care, and so on – they also become proletarian. Exploitation of the general intellect means that these resources, exhaustible at the level of the individual worker, become abundant. This *proletarianisation of the social* feeds the precariousness of cognitive labour, as there is always a reserve to draw on when the individual burns out. This perverse flexibility undermines the status of the worker, who can scarcely feel confident about their role when they are so readily replaceable. Short-term is widespread, workers are unable to adequately plan their futures because they are uncertain, and so cannot project themselves meaningfully beyond a miserable present. Where work might once have grounded a sense of identity (Bauman 2004), its precariousness now fosters anxiety. As Paolo Virno (1996: 17) argues, no longer is it fear of privation that causes us to seek out work, there finding security and stability, but instead, that the conditions of work today produce such anxieties. Precariousness is not just a condition of cognitive labour in communicative capitalism; it is also a mental malady. Richard Sennett (1999: 70) argued in the late 1990s that people no longer cared about the work they were doing and that, as such, they could not feel alienated by their labour; in the place of alienation was now, simply, indifference. More plausibly, Virno (1996: 15) has argued that precarious and cognitive labour do not so much cause alienation as reduce workers' experience of it to a job description. Mental illness is one of the forms this alienation takes. The intensification of labour, in a bid to increase productivity, most notably accelerated by information and communication technologies, simultaneously produces 'misery, the subordination of human beings to wage labor, solitude, unhappiness and psychopathology' (Berardi 2009a: 58). The conditions of work themselves are pathological, and being constrained into adjusting to the hyper-interactive, over-stimulated, dangerously competitive and ultimately precarious environment results in psychopathology. Mark Fisher (2009: 19) has argued that the 'mental health plague' afflicting advanced economies shows capitalism to be inherently dysfunctional, with a high cost for those that participate in it. Depression, anxiety, insomnia, attention disorder and stress are included in the job description. Employing his concept of *capitalist realism* – that capitalism appears to us now as a natural order, such that any disorder, any crisis, is experienced as akin to a natural disaster like an earthquake or a hurricane, rather than an avoidable pathology of a socially constructed system – Fisher (2009: 19) argues that we lose sight of the connection between our socioeconomic order and the mental disorder inherent to it. Mental health issues are seen to be private affairs, personal tragedies that are never politicised, with all the explanatory weight carried by reference to brain chemistry and never to social systemic factors. As Fisher (2009: 37) observes, chemico-biology can of course explain the reactions in the brain that instantiate illnesses such as depression, but this narrow focus may miss the external factors that cause them, concluding that 'repoliticizing mental illness' is 'urgent' if we want to challenge the conditions that destabilise mental equilibrium and mask the very processes by which this is achieved. If not, then we face a crisis of over-production of nervous-exhaust.

Berardi (2009a: 42) summarises this coming crisis in the starkest of words: 'Today capital needs mental energies, psychic energies. And these are exactly the capacities that are fucking up'. But if critics have been slow to politicise mental health issues, then the pharmaceutical corporations have had a much better grasp of the structural problem. The 1990s saw rising use of cocaine and amphetamine, particularly amongst so-called creative labourers (Berardi 2012: 98); to keep up with the hyper-stimulation of semiotic acceleration, what better drug to self-medicate with than speed, which is scarcely any different from the legal treatment for ADHD? At the opposite end of the spectrum is heroin, rising in popularity in the 1980s and firmly held there by the grunge-glamour of Kurt Cobain and Layne Staley in the decade after, a drug that slows down, that allows for disconnection. The choice here is one between speeding-up and slowing-down, keeping pace with the insane duration and acceleration of communicative capitalism, or dropping out. The pharmaceutical companies are on hand to make sure workers are able to keep up. Take, for example, the treatment of depression. Diazepam (Valium) leads to sedation and amnesia, general relaxation, whilst SSRIs such as fluoxetine

DOI: 10.1057/9781137394781.0004

(Prozac) work to increase serotonin levels; as such, it is not surprising that the latter have largely phased out the former, since they allow for a euphoric and energised workforce, rather than a relaxed and sedated one (Berardi 2009a: 121). Or take the following examples: when sleep becomes a scarce resource it becomes possible to charge people for it in the form of sleeping pills (Crary 2013: 18); or when there is an over-production of attention it is possible to prescribe psycho-stimulants such as methylphenidate (Ritalin or Concerta) in order to enhance alertness and general mental acuity, and so increase performance in the work-place. Federico Campagna (2013: 12) writes of an 'army of the tragically overworked' that is 'fed on psychoactive drugs and self-help remedies'. He argues that work is the new religion, and that religion was never the opium of the masses, since opiates are calming, but rather more like amphetamine – perhaps more so like crystal meth – making the masses hyperactive (2013: 28). When labour moves from being primarily physi-cal to primarily cognitive, mental health issues come to the fore; where previously a depressed worker was of little concern, so long as they kept their body healthy, today mental illness needs to be met by pharmaceuti-cal corporations in order to sustain a sick but productive workforce. But eventually this development will chase itself off a cliff; there is only so long people can engage in work that is fucking them up, regardless of the chemical fix.

When this happens, when Campagna's army of over-worked and drugged-up labour finds itself replaced by the reserve army of under-worked but no less precarious labour, and falls into what was once the safety net of the welfare system, they become prone to the social experi-mentation and control of governments. Standing (2013: 142) observes the development of a 'therapy state' in response to the recent economic crisis: 'In the United Kingdom, after the shock of 2008, instead of dealing with the structural causes of stress and depression, the government mobilised CBT (Cognitive Behavioural Therapy) to treat the outcomes'. What we get, instead of any serious attempt to understand unemployment, indebt-edness, reduction in real-terms wages and so on, as part of an economic system that has suffered systemic failure, is that only the symptoms are treated, and so a focus on improving access to CBT through the NHS and mental health co-ordinators at Job Centres. 'Instead of recognising the causes of difficulties', writes Standing (2013: 142), 'the intention was to treat the victims of economic mismanagement and encourage them to think they needed therapy'. Anxiety and depression are understood here

DOI: 10.1057/9781137394781.0004

as individual abnormalities, when really they are pretty normal responses to unemployment, dire economic prospects and general precariousness. 'Soon', predicts Standing (2013: 142), 'the powers that be will be saying that, unless people take a CBT course, they will lose entitlement to benefits'. It did not take long. In August 2014, the *Guardian* (Marsters 2014) reported a scheme by the Conservative-Liberal Democrat coalition government in the UK to tie Employment Support Allowance (ESA) to Cognitive Behavioural Therapy. ESA is a benefit provided to people who are too ill or disabled to work, and the scheme, at the time a voluntary trial but with plans to make it mandatory across the board, involved those claiming ESA with diagnosed mental health issues only receiving their benefits if they attended CBT schemes. Helping people to get better is one thing; threatening to immiserate them by cutting off welfare payments if they do not take the government approved route to getting better is quite something else.

Both pharmaceutical corporations and governments recognise that mental health is as much a structural problem as it is individual: the drug companies sustain productivity in the face of forms of labour that exhaust mental energies; and the government tells some of its most vulnerable people that they are sick and need therapy, whilst trying to force them off a welfare bill that they cannot afford because of a dangerously volatile economic system that reduces many to anxiety, depression and restlessness in the first place. The psychopathological development of communicative capitalism is an urgent site of resistance. However, the danger is that these conditions foster not only psychopathology but also sociopathology, that precariousness – of labour, of life, of mental stability – provides unsuitable grounds for community and solidarity, and that cognitive labour, and the communication technologies that make it so profitable, are *dis-empathising*. Jean Baudrillard (2011: 15) once wrote that the 'subject deprived of all otherness collapses into itself, and sinks into autism'. The next chapter will address the issue of communicative capitalism in terms of its social pathology, that is, how the technological organisation of both our work and leisure time minimises the other (person) whilst maximising their transactional presence on a screen and fostering a social anxiety borne in disconnection and the instrumentalisation of language.

DOI: 10.1057/9781137394781.0004

3

Social Anxiety

Abstract: *This chapter explores the contactless contact of our productive social media interactions alongside the isolation and fragmentation of cognitive labour. It is argued that social media has allowed us to migrate interactions to screens such that we can maintain contact without the spontaneity and risk of proximate encounters. At the same time, remote working and alternative officing practises constrain workers into forms of networked communication that lack the thickness required for empathy and solidarity, stripped of the unsaid and the ambiguous in order to allow for efficient exchange of brute information. The sort of certainty and control we can achieve in social interactions originate from the same source as our precariousness and loss of autonomy at work: the circuits of* communicative capitalism.

Hill, David W. *The Pathology of Communicative Capitalism*. Basingstoke: Palgrave Macmillan, 2015. DOI: 10.1057/9781137394781.0005.

DOI: 10.1057/9781137394781.0005

In 1985, thirty-nine football supporters lost their lives after crowd trouble during the European Cup Final contested between Liverpool and Juventus. In *The Transparency of Evil: Essays on Extreme Phenomena*, Jean Baudrillard reflected that violence in the stands was an attempt by the spectators to transform themselves into actors, a reversal of roles whereby 'under the gaze of the media, they invent their spectacle' (Baudrillard 2009: 87), and the non-event of a game abandoned in spirit, if not in reality, is played out across the world's screens with its focus firmly on these actor-spectators. Baudrillard's essay is not so much concerned with understanding the Heysel Stadium Disaster as an event, but in situating this explosion of violence – of hooliganism that led to football fans being crushed to death under a collapsed stadium wall – within the context of the screen and of an already emerging participatory culture, or interactivity. Baudrillard asks:

> Now is this not precisely what is expected of the modern spectator? Is he not supposed to abandon his spectatorish inertia and intervene in the spectacle himself? Surely this is the leitmotif of the entire culture of participation? (2009: 87)

For Baudrillard (2009: 90), Heysel was participation taken to 'its tragic limit', but it was nonetheless consonant with a then emerging, and soon to be fully formed, injunction to foreswear the passive consumption of media, to act – to interact. What is more, the example of football violence takes us further into an understanding of the nature of the Baudrillardian non-event, soon after to be exemplified by the First Gulf War (Baudrillard 1995). Teams found guilty of crowd trouble can be forced to play future matches behind closed doors, the stadium empty but the match nonetheless broadcast for a television audience. This, says Baudrillard (2009: 90), 'exemplifies the terroristic hyperrealism of our world, a world where a "real" event occurs in a vacuum, stripped of its context and visible only from afar, televisually'. Without the presence of fans, what occurs is an antiseptic media spectacle, always already an image divorced from immediate sensation:

> Here we have a sort of surgically accurate prefiguration of the events of our future: events so minimal that they might well not need take place at all – along with their maximal enlargement on our screens. No one will have directly experienced the actual course of such happenings, but everyone will have received an image of them. A pure event, in other words, devoid of any reference in nature, and readily susceptible to replacement by synthetic images. (Baudrillard 2009: 90–91)

DOI: 10.1057/9781137394781.0005

It would be possible here to argue that Baudrillard had predicted the later rise and now domination of Sky television in the football broadcasting arena, its principal product, the Premier League, beamed to the screens of millions around the world whilst relatively few (can afford to) experience matches in football arenas across England and Wales. Of more interest is this early account of media interactivity. In opening his essay, Baudrillard describes the Heysel Stadium Disaster as a 'simulacrum of violence', whose origin for those who bear witness televisually was not so much in passion but of the screen, a kind of terror that 'exists potentially in the emptiness of the screen, in the hole the screen opens in the mental universe' (Baudrillard 2009: 85). Then, in conclusion, he writes: 'Every real referent must disappear so that the event may become acceptable on television's mental screen' (Baudrillard 2009: 91). What happens when media interfaces become so prominent to both our work and social time? What is the impact of days spent inputting, interpreting and/or communicating information in the pursuit of a salary and in the maintenance of interactions with others? The argument in this chapter is that the centrality of the screen, as both window and mirror, breeds a kind of social anxiety, the cause of which is the marginalisation of the other as a real referent – a cause which has been perversely mistaken for the means to manage this anxious disposition to social existence and proximity to others.

In 2012, Amazon launched a television advert with the slogan: 'Connecting your mouse to your front door was our moon landing' (see YouTube 2012). The message was clear: the negation of place is the new space race. It was an advert that brought to mind the writings of Paul Virilio. Take, for example, this remark made in conversation with the media theorist Friedrich Kittler:

> For me, the new technologies make space disappear into a void, in its extent and its time. This is a profound loss, whether one acknowledges it or not. There is also a pollution of the distances and time stretches that hitherto allowed one to live in one place and to have relationships with other people via face-to-face contact. (Virilio & Kittler 2001: 102)

This line of argument is perhaps best elucidated in his book *Open Sky* (1998), the title of which speaks to exactly the kind of shift hinted at by Amazon's space race. Virilio (1998: 2) argues that we have become so obsessed with a time race, with increased connection speeds, real-time transmission, instant interactivity and so on, that we lose sight of the

DOI: 10.1057/9781137394781.0005

implications. The space race, lest it be forgotten, meant leaving behind the real-space of the planet of human inhabitation, and, similarly for Virilio, an obsession with the speed of media and communication involves a negative relationship with the immediate environment. It is by looking up to the open sky, he says, that we are reminded that technologies may cost us the world. Communication technologies allow users to interact at the site of an interface machine, giving rise to what Virilio calls 'terminal-man', who controls his interactions at the screen of the terminal (but who is also *terminal*, one might suspect, in the sense of coming to an end) and whose interactivity through networked technologies is ultimately a kind of inertia (1998: 10–11; 16). Provocatively, he suggests that the terminal-man – in its simplest form, users of communications technologies such as the Internet – takes as his inspiration 'the pathological model of the "spastic"' (1998: 20). Virilio here seems to be referring to the paraplegic, who, despite having limited mobility, has technological assistance in order to navigate and control their environment. His terminal-man, then, uses online interaction and the like in order to remotely control their encounters with others. This kind of remote control can be understood as the ease of starting and terminating interactions online, the possibility to delete or block users whose attentions are undesired, the ability to carefully construct and curate social networks and so on, all whilst keeping others at a distance. However, Virilio (1998: 19–20) takes his argument further – perhaps too far – and argues that people now favour mediated interactions over the face-to-face and, what is more, encounters with people who are remote rather than proximate, 'the distant' (or stranger) over 'the near (and dear)', making strangers of people who are local in order to interact with people online who are far away in a wholesale 'inversion of social practices'. This would be a reversion to the law of least action, since there is a connection between space and effort; going out and meeting people when you can log-in and meet anyone through a computer, tablet or smart-phone is like taking the stairs when you could ride the escalator. As such, argues Virilio (1998: 62), 'the *microphysical* proximity of interactive telecommunication will surely see us staying away in droves, not being there anymore for anyone, locked up, as we shall be, in a *geophysical* environment reduced to less than nothing'.

That networked communications indicate an inclination for interaction with strangers over proximate others *may* have sounded plausible in the 1990s, the time in which Virilio was writing, but it does not ring

DOI: 10.1057/9781137394781.0005

true today given the nature of social networking sites which, as Jodi Dean (2010: 58) observes – here of Facebook – are popular because there are few strangers with direct access to users. It is also worth remembering that communication is not a zero-sum game, that it is possible to increase online communication without a balancing decrease in co-present interaction, and that, if it were possible to categorise contacts into online and offline, there would be a huge crossover in group membership. Text messages, social networking invites, instant messaging, webcamming and so on, allow users to negotiate social interactions with those near (and dear). That said, even if Virilio's claims about strangers lack plausibility, it is possible to observe a kind of distancing that can be facilitated by these technologies – again, a sort of remote control. Sherry Turkle (2011), for example, notices an aversion to intrude on each other in real time and space, such that a bombardment of text messages or emails is preferred, so totally intruding on the other electronically; she observes that our connectivity with others offers the illusion of companionship without the demands of friendship, essentially meaning we engage in looser (proximate) relationships that we can control technologically without feeling lonely; and that, ultimately we keep in touch with people that we also keep at bay. This is the fate of our colleagues, our casual acquaintances – maybe even our friends, reduced to contacts that we may rarely have physical contact with. It might be, as Jonathan Crary (2013: 59–60) suggests, that, because of the infinite content and interactions they open up, network technologies provide access to information and people that will always outdo what is immediately around us, will always trump naked human communication. This is a sort of consumerist's wager – Why put all your eggs in one basket with a co-present encounter when you can hedge your bets online? – which is entirely plausible, but it does not explain the sorts of scenarios Turkle describes since they exist within already existing friendship groups. Crary (2013: 124) comes closer to addressing these issues when he writes that the 'responsibility for other people that proximity entails can now easily be bypassed by the electronic management of one's daily routines and contacts'. Slavoj Žižek (1999) once pondered on the appeal of the Tamagotchi, a late-1990s Japanese toy craze where digital pets were raised on a key-ring-sized, egg-shaped computer, fed on virtual food and human attention, or else neglected and ailing – and ultimately dead. Why do we invest time in nurturing something that ultimately is of no real consequence? The answer is perhaps that humans have a moral duty to care for others and something like a Tamagotchi

DOI: 10.1057/9781137394781.0005

allows us to fulfil this duty by caring for a synthetic other – without all the bother of having a real person on our hands. The digital pet allows for a programmed routine of social care and responsibility whereas other people are altogether less predictable. Social life is a mess of contingency, of random encounters, and our social obligations often appear as interruptions without warning. Ironically, as we come under ever greater pressures to be available to communication, so escalating the ways in which we can be made aware of our responsibilities to others, these technologies bring in the element of control afforded by the Tamagotchi, allowing us to switch off, ignore or delay, which is why Žižek (1999) concludes that both Japanese toy and mediated encounters reduce the other (person) to an interface simulacrum, something altogether less demanding.

One of the most striking examples of this remote control of social encounters and management of responsibilities is the contemporary and mediated form of stalking. Zygmunt Bauman (2008: 93) has suggested that the fear of stalking is a sign of the times, arguing that stalkers act as modern-day bogeymen when societies become fragmented. However, a kind of stalking that aims at the remote control of encounters and that takes place through social networking sites such as Facebook has shown that it is now not so much a bogeyman but a way of managing the anxiety of proximity to others. Something like Facebook stalking is not 'real' stalking – the threatening harassment of another – by other means; rather, it relates to a set of new practices, made possible by social networking sites, that aim at discovering information about people that would not normally be available to the stalker. Facebook users define it simply as 'browsing' the profiles of other users, and suggest that it is a normal part of Facebook use (see Lewis & West 2009: 1215). We might better understand what is happening here by way of Žižek (2008: 35), who argues that the 'liberal tolerance towards others' we see today, 'the respect of otherness and openness towards it, is counterpointed by an obsessive fear of harassment'. He observes that 'in today's ideological space, very real forms of harassment such as rape' – and we can add stalking – 'are intertwined with the narcissistic notion of the individual who experiences the close proximity of others as an intrusion into his or her private space' (2010: 5). Our aversion to over-proximity has two consequences, the first of which is that we can only tolerate others if they keep their distance:

> Tolerance means: no harassment. Harassment is a key word. Fundamentally, what this says is: hide your desire; don't come too close to me. ... That shows

DOI: 10.1057/9781137394781.0005

us that tolerance in this context is precisely a form of intolerance: intolerance for the closeness of the other. (Žižek in Badiou & Žižek 2009: 92–93)

The second consequence is that we are ourselves duty-bound to keep our distance from others, that tolerance of others goes hand in hand with an injunction not to intrude on others, not to get too close to others or invade their space (Žižek 2008: 35). Žižek illustrates this in typical style: 'It means, as I have experienced in the US: If you look too long at somebody, a woman or whoever – that is already a visual harassment; if you say something dirty – that is already verbal rape' (in Badiou & Žižek 2009: 93). As a result, we have seen not only much needed legal injunctions against harassment, stalking and other such possessive behaviours, but also social and moral regulation of body-space and public attention, which create uncertainty around social interaction and concern about over-proximity, and stalking through social networking should be considered as a coherent but ultimately forlorn attempt to manage this. Stalking is not an experience common to the majority of people, even if the fear of stalkers is; online, though, we can all be drawn in, compelled to a virtual over-proximity. Social network stalking allows us to satisfy the voyeuristic urge whilst keeping a distance, so avoiding the burdens of sociality and responsibility, and, at the same time, fulfilling our obligation not to encroach on others, to keep our distance. Žižek can look all he wants at women without being accused of visual harassment so long as he signs up to a social networking site. In such an environment, the other person becomes mere pornography (Berardi 2009b: 172), a virtual contact without any social encounter or *quid pro quo*, enjoyed from afar. Social network stalking – knowledge of the other from a place of hiding – should be understood as an attempt at the remote control of the contingency of social existence. But in the end it is unlikely to restore confidence in proximate encounters.

Social networking technologies and mobile media are leading us towards a reconceptualisation of what it means to be in contact:

> The very word "contact" comes to mean the exact opposite of contact: not bodily touch, not epidermic perception of the sensuous presence of the other, but purely intellectual intentionality, virtual cognizability of the other. (Berardi 2012: 111)

By way of example, in 2012 it was announced that researchers at MIT had developed a vest that gave the wearer a 'hug' (really a contraction of the garment) every time something they posted on Facebook received a

DOI: 10.1057/9781137394781.0005

'like', that is, was rated positive by someone in their social network (see Wainwright 2012). The vest was not intended to be commercially viable – it was developed in order to explore the potential of haptic systems – but it works as a satire of the contactless contact that occurs through social media, where the tactile contact would have to be artificially reinstated in order to properly flesh out interaction with others. What is curious about social media like Facebook is that they allow users to cultivate a social network such that they can be freer – less anxious – in their self-expression since, largely, such a group consists of friends and acquaintances rather than strangers, whilst at the same time, users are encouraged to expand that network endlessly, tempted into becoming over-exposed and losing control (Dean 2010: 65). That the owners of these sites would wish to encourage this, prompting users with recommendations for new friends based on friends of friends and so on, is hardly surprising given the value of this sort of relational data, but the effect is that sociality is held in abeyance. The bloated social network may now include people with whom encounters are yet to come, or, the promise of encounters that is never fulfilled. What users end up with, then, is 'friendship without friendship' (Dean 2010: 35), comfort in the knowledge that they are not alone, this connectedness maximised on the screen but minimised in its fully sensuous, intimately tactile, emotionally involved togetherness. What we get is a sort of *companionship-in-hiding*. As Virilio (1998: 26) argues, everything loses its weight when translated into an image on a screen. People become less dense, less of a burden – or a burden more easily lifted. People dematerialised are easily dispersed into the ether at the click of a button. And if people simply do not want to be alone, whilst at the same time protecting themselves against the contingency of sociality and the burden of responsibility, then the screen, with its effacement of the near and the far and its denaturing of the encounter, with all its contingent risk, becomes a useful aid. The screen – of the computer, of the smart phone or the tablet, the smart watch and other mobile and wearable media – becomes a prophylactic to social anxiety. The problem here is not some fetish for the strange, nor simply a law of least effort, but the anxiety occasioned by proximity and the individualistic social norms that have come to govern it. To return to Baudrillard, but considering here sociality instead of football violence, the screen has become the mirror of our social anxiety. His account of interactivity creating actor-spectators is not contradicted by the passivity of the likes of the social network stalker. The actor-spectator is a product of the way that interactivity combines interaction with inaction; where the football fan shook off their 'spectatorish inertia' to act in the event, the social actor now adds the very same thing to the event of sociality, to the encounter. Social interaction plays out in empty stadiums, stripped of thick communicational context – tactility, sensuousness – televisual from afar in the vacuum of the screen. The social relationship becomes so minimal that it need not take *place* at all, at the same time that we see its maximal enlargement on our screens. Direct experience gives way to reception of an image or a shadow that need not necessarily have any referent beyond our communication devices. Baudrillard argues that the screen opens a hole in the mental universe, the possibility of terror, but that if all real referents disappear, then the televised non-event becomes acceptable. Social anxiety is a result of an inability to cope with the demands of intersubjective encounters; social media is the possibility of infinite encounters, and therefore, the potential for anxiety to colonise our thinking about others – but it also allows for the erosion of real referents, for social relationships to play out on the screen, its sterility affording a sense of security.

'Non-stop interactivity', writes Guy Standing (2013: 131), 'is the opium of the precariat, just as beer drinking and gin drinking was for the first generation of the industrial proletariat'. It is instructive to bear in mind the connection between the forms of communication that take place in social relationships and those that generate wealth through cognitive labour. The isolated user of social media and communications technology is doubled by the increasingly isolated worker in a flexible economy (Dean 2009: 4). For example, the remote control of the social is carried through to interactions with service workers, as when ears are turned inwards by headphones as we board the bus, past the driver, or when shopping is checked out at the supermarket, mobile phone acting as an im/material barrier to the assistant. The checkout assistant in particular demonstrates not only our social anxiety, but also the precarious balance between cognitive labour as Human Intelligence Tasks and automation. It is a job that requires small talk, appearing friendly, being generally cheery – all of which have today become employable skills (Standing 2013: 123). Turkle (in Turkle & Nolan 2012: 63–64) worries that automated checkouts will have a negative social consequence, in that once shoppers see that a machine can do the job of a human, they might, when using a staffed checkout, treat the human like a machine. This could take the form of a general discourteousness, of failing to end a phone conversation or to switch off one's music or to stop emailing whilst being served, so tuning out the social skills the checkout assistant has to put to work to make a living. More to the point, though, the automated checkout reveals that, at its base, this is a job that can be automated and as such is not a Human Intelligence Task, in the sense of a function that cannot be performed by a computer. Here the shopper is brought into a confrontation with their own potential redundancy through automation. It reminds us that flexibility has come at the expense of employment stability, that is, that automation has the ascendency over autonomy. This is also, in part, a result of (1) alternative office practices (such as remote working) and (2) the kinds of communication that occur in the workplace. As will be seen, the combination of these two factors risks a diminishment of empathy and the erosion of solidarity.

In order to fully understand what is at stake here, it is important to clarify the moral value of empathy before contextualising its diminution within the present precariousness of labour and the effect this has on community and solidarity, something best achieved by contrasting empathy with sympathy. The most enduring accounts of sympathy can be found in the classical liberal philosophy of Edmund Burke and Adam Smith (see Eagleton 2009: 62–82). For Burke (2008), communities are bound together by mutuality and affinity – resemblances, conformities, customs, and habits – that allow individuals to engage in imaginative sympathy, that is, to put themselves in the other's shoes and feel responsible for them. This can only work in proximity and likeness, such that people who are unalike fall beyond sympathy. The moral transitivism of Smith (2009) is similar, insofar as sympathy is grounded in an imaginative act of recreating in the mind the condition of a person encountered, but he rejects the parochialism of Burke, arguing instead that both the conduct and conditions of those outside of one's community cannot be justified by difference where they are unjust or unreasonable. The disagreement here is ultimately indicative of the inappropriateness of sympathy as a moral concept: sympathy requires not only understanding but something shared, it would aim to pick out sameness, which results either in narrowing sameness to a community group (Burke) or expanding some slight sameness universally (Smith); and sympathy not only operates on understanding but also through pity, which means there needs to be some imaginative leap in which the sympathiser assumes the pain of the other. Empathy is a much more useful sentiment. Empathy

DOI: 10.1057/9781137394781.0005

is openness to difference and receptivity to the communication of pain. This distinction is well-drawn by Richard Sennett:

> Both sympathy and empathy convey recognition, and both forge a bond, but the one is an embrace, the other is an encounter. Sympathy overcomes differences through imaginative acts of identification; empathy attends to another person on his or her own terms. Sympathy has usually been thought a stronger sentiment than empathy, because 'I feel your pain' puts the stress on what I feel; it activates one's ego. Empathy is a more demanding exercise, at least in listening; the listener has to get outside him- or herself. (Sennett 2012: 21)

Empathy is a craft of understanding and responding to other people. It requires attentive communication, listening to others, and responding to the other person such that communication progresses whilst keeping the differences between interlocutors intact, so constituting a meaningful encounter since the other person is met on his or her own terms. Is there any time left for this kind of empathetic communication? Is there any space available?

If communities in an industrial age were formed around the workplace – the mine or the factory – such that people who worked together, lived together (Jones 2012: 142–43), and if this situation led to greater strength, social cohesion and empathetic bonds (Harvey 2012), then an important question must be raised: at what price have we been sold fragmentation and flexibility? Community is more than just communication, it is a form of sharing (Crary 2013: 119–20), and in order for communication to embody sharing, it cannot consist solely of transmission of information and pared-down messages. We have to share meaning in order to be able to ground empathetic understanding. This is not best achieved through abstracted and immaterial interaction but by sharing sites and routines – and yet this kind of synchronised and repetitive contact runs against the business mantra of flexibility. For example, 'alternative officing', the business practice of having workers frequently change the location they occupy, is designed to add spontaneity to the work day in order to facilitate creative new collaborations. This incorporates strategies such as 'remote working'; 'hot desking', where there are more employees than desks such that workspaces are shared asynchronously; 'hoteling', where workspace is on demand or just in time; or 'activity-based working', where there is no set workplace and the site of employment is determined according to the kind of work that needs to take place (see Humphry 2014: 359–60) – all of which reduce costs for the

DOI: 10.1057/9781137394781.0005

business, but increase the social cost to the employees. No longer is there a community of workers who share a common purpose, a site of labour, and a work temporality. Workers become isolated, no longer together but alone in cubicles, forced to wait for a colleague to finish before being able to occupy a desk or with no fixed work environment at all, and constrained into shallow team-working with an ever-changing cast of temporary colleagues. Work, then, becomes less predictable and more uncertain. Workers are deprived of the security of their surrounding social and material environments of action even though shared space is necessary to ground solidarity and worker autonomy. Instead, the world of work is altogether more precarious. We might yearn for community, as Sennett (1999) concludes in his exploration of work conditions in the late 1990s, but short-termism and the demand to be flexible and mobile are infertile grounds for developing social bonds – something precarious workers may even become wary of doing, since they will rarely last (Standing 2013: 22). Career-less work, fragmental labour, is insufficient to allow the workers to generate a work identity that would lead them to identify with a work community, since they cannot tap into the social memory, traditions and norms of the job over time (Standing 2013: 12). Precariousness is not only about a lack of job security and flexible work conditions – it also confers a loss of collective identity:

> Precarity refers not only to the deregulation of the labor market and the fragmentation of work, but also the dissolution of community. A continuous flow of infolabor runs in the global network, and it is the general factor of capital valorization, but this flow isn't able to subjectivize, to coagulate in the conscious action of the collective body. (Berardi 2011: 129)

Bauman (in Bauman & Donskis 2013: 64) suggests that the precariat – which he identifies with 'the 99%' of Occupy slogan fame – are fearful of and disgusted by other people in the precariat, essentially people like them. This represents a catastrophic failure of empathy, which is perhaps inevitable in a world of work organised around the mantra of 'No Long Term', where working relationships are short lived and precarious workers are in constant competition with one another.

Communication and flows of information are essential for the flexible work practices that spread precariousness, allowing for the break-up of the rigid working conditions of employees and the loss of the shared space of labour. The co-worker becomes part of a system of signs, communication taking place at an abstracted, informational level. As Berardi

(2009a: 70) argues: 'The experience of the other is rendered banal; the other becomes part of an uninterrupted and frenetic stimulus, and loses its singularity and intensity – it loses its beauty'. Communication with others becomes lost in the overstimulation of information, becoming less human and altogether more functional, flowing without uniting. Cognitive labour extracts value from human subjectivity by putting to work the affective attributes of the worker, amongst them language, expressiveness, creativity and emotional connection; under the conditions of communicative capitalism these qualities of linguistic practice are measured only in terms of their instrumental worth, such that they are directed outwards to customers (call centres, service work, etc.) but regarded as of little value within a workforce, among colleagues. Instead, communication at work is subjected to the logic of efficiency and exactness, with elements that do not pull in this direction – polyvocality, multi-dimensionality, ambiguity and so on – neutralised; as Lazzarato argues: 'Neo-capitalism asserts the primacy of languages of clarity, precision, functionality, and instrumental and pragmatic efficiency by vacating them of the expressive dimension of humanist languages' (2014: 73; 130). Whilst it is true that workers can carve out spaces for more expressive communication – breaks, face-to-face meetings and so on – when productive communication is migrated to enclosures of code then the time for such communication falls under a strict regime of control. Workers receive an uneven access to ordinary talk, leading to 'communication deskilling' (Boden & Molotch 2004: 104–05) by operating in a work environment that strips language bare. Ordinary talk is rich in informational cues such as facial gestures, body language, the intonation of the voice, eye contact, tactility and emotional expression, but these slow down the exchange and, redirected through communication technologies, talk is reduced to brute information such that work rarely offers the opportunity for affective bonding with co-workers at the same time that what is communicated loses so much of its meaningful, human dimension. Communicative capitalism ensures an increase in circulation of information but, without expressiveness and the affective power this confers, also less meaning, since this requires time for reflection on the unspoken and the ambiguous, which slows down exchange. Meaningful human communication becomes a victim to speed. Dismantling the human environment of language in order to reduce communication to an efficient, profitable exchange deskills us in sensibility – the ability of humans to communicate what cannot be said with words – and

DOI: 10.1057/9781137394781.0005

encourages instead a sensitivity to information and data (Berardi 2015: 49). Workers are no longer afforded the time or the space to be receptive to the bodily, emotional expression of their colleagues such that the kind of nuanced understanding of the other that is essential to Sennett's conceptualisation of empathetic communication goes missing. And without being able to fully share experiences, hopes, suffering, fears and so on, workers are denied the opportunity to reflect on how they share in those feelings. Without the opportunity for more empathetic forms of communication – thick with information more valuable socially than that which is valued financially – isolation thrives over solidarity, and individualistic competition between workers is encouraged at the cost of community resistance against shared precariousness. As Berardi (2015: 171) observes: 'Loneliness and the lack of concern for fellow workers are simultaneously the cause and effect of the absence of collective action'.

According to the logic of communicative capitalism it is desirable for us to divert our energies into social media, since this generates value. At the World Economic Forum in Davos in 2015, Facebook claimed that it had a 'global economic impact' of $227 billion over the previous year, as well as directly and indirectly generating 4.5 million jobs, figures that John Naughton (2015) described as 'bullshit'. Bullshit they may be, but the perception of value is all that really matters. Users are sold social networking in terms of the efficiency of managing their social existence, but the real efficiency comes in the uploading and exchanging of personal information in a way that can be processed with maximal operativity. We seek security from the spontaneity of social existence but the remote control of others through social networking technologies is a perverse doubling of the remote control exerted through the contemporary workplace and the practices of cognitive labour. The rationalisation of language found in the workplace is also apparent in our text messages and social networking posts, where vowels are a luxury, and in the character constraints of micro-blogging sites such as Twitter (Bauman & Donskis 2013: 46), where communication takes place at the risk of communicability (Dean 2012: 127), that is, where meaning gives way to a bare sensitivity or receptiveness to the purely syntactical. The intersubjective security that is perversely desirable – since sociality is a contingent, unpredictable event – is provided by the same source that makes the environment of work so precarious. Communicative capitalism provides the means for etherealising the burden of others, whilst removing from our grasp those we might lean on to fight for better

DOI: 10.1057/9781137394781.0005

conditions in the workplace. Networked communications, in work and leisure, reveal themselves to be more conducive to undoing bonds than reinforcing them, realising Tiziana Terranova's (2004: 145) warning that communication in itself is no guarantee of community. Reducing the presence of the other to the screen only appears to be a solution to social anxiety because we have witnessed a wholesale deskilling of empathetic communication, and by minimising encounters with others we reduce the chances of developing the solidarity needed to challenge the pathological development of communicative capitalism.

DOI: 10.1057/9781137394781.0005

4
Pathological Development

Abstract: *This chapter explores the idea that technological development is inherently pathological as long as it is co-opted by* communicative capitalism. *The arguments of the preceding chapters are returned to through this framework, whilst debt is introduced as one of the biggest threats both to mental health and to imagining a different future. The aims of this chapter are modest: it concludes the book by summarising the key sites for resistance – both external and within left politics – and repeating the call for* autonomy over automation; *no manifesto is offered since how this translates into action is an empirical affair beyond the scope of this theoretical exploration.*

Hill, David W. *The Pathology of Communicative Capitalism*. Basingstoke: Palgrave Macmillan, 2015. DOI: 10.1057/9781137394781.0006.

In one of the provocative essays on time and technology collected in *The Inhuman*, Jean-François Lyotard (2004: 8–23), reflecting that the development of an information society is a simple matter of achieving greater efficiency, argues that the only limit to this would be the heat-death of our sun and the resultant destruction of the solar system. About 4.5 billion years old, the sun is in its middle age, adding urgency, he teases, to the whole affair. If such a heat-death poses a limit to development – which would otherwise be without finality, since its only operating principle is internal improvement – then the challenge for the system, that is, the combined and complicit forces of the sciences, technological innovation and capitalism, is to find a way around this limitation. With the death of the sun comes the destruction of all humanity, along with development, which Lyotard characterises as a parasite on the human host. To survive this, there needs to be found a way for humans to exist outside the solar system, and this job is already underway – it is just that we humans are not knowingly driving it. According to Lyotard, in the technical and scientific research of *all* fields, regardless of what the immediate goal is said to be – war, health, communication, whatever – the ultimate aim is to make existence possible after the death of the sun. Now, the technological sciences understand the human in technological terms: it possesses hardware, the body, which sustains the capacity for thought; and software, language, which makes thought possible. Of course, the hardware will be consumed in the solar explosion, and so too thought. As such, the technological sciences face the problem of designing hardware that can support this software in another environment to that of earth, that is, they face the problem of making thought possible without a body: 'But "without a body" in this exact sense: without the complex living terrestrial organism known as the human body. Not without hardware, obviously' (Lyotard 2004: 14). It falls to roboticists or computer engineers to construct a vessel for thought, and it falls to artificial intelligence research to create thinking software programs to imbue these machines with an approximation of human intelligence. In this sense, development can continue beyond the heat-death of the sun and without a human host. But sadly, Lyotard concludes, the human race has had it.

It would be easy to dismiss this quasi-science-fictional account as mere provocation, or playfulness. However, it might instead be read as a fable – a story with a moral – that at the same time illustrates and underwrites Lyotard's entire critique of progress. The over-arching narrative of this

DOI: 10.1057/9781137394781.0006

story is that advanced capitalism seeks only to advance further. As such, it must become more efficient, which involves saving time, efficiency being the driving force of development. In *The Postmodern Condition* Lyotard (2005) introduced the concept of performativity, which he argues is the operating principle of what he calls 'techno-scientific capitalism', that is, an economic system that advances itself through the complicity of technological and scientific research in order to tighten its grip on society and reap higher dividends. Performativity is a principle of efficiency, of achieving maximal output for minimal input, optimising performance within the system (Lyotard 2005: 11). One of the most lucrative ways of ensuring efficiency is the introduction of time-saving technologies, in particular, information communication technologies. Information is frictionless, stripped bare of humanist meaning, exchanged profitably with maximal efficiency. Capitalist development is ensured by technological revolution, but this has nothing whatsoever to do with Enlightenment ideas of progress, no consideration of what is for the good of humanity, and is without any movement towards truth or justice; this is a purely technological movement co-opted to reinforce and tighten the system's grip on society. Technologies follow the law of least effort, working to a principle of optimal performance by maximising outputs and minimising inputs, that is, to generate or capture as much information as possible for the least amount of energy expended. Under this regime of control, the very idea of what is *good* is transvaluated; what is *better* is a technological movement that generates the same value or more with less effort (Lyotard 2005: 44). Development is an end in itself, striving only to achieve higher performance/efficiency and greater profits. As Lyotard argues:

> It is no longer possible to call development progress. It seems to proceed of its own accord, with a force, an autonomous motoricity that is independent of ourselves. It does not answer to demands issuing from man's needs. (1992: 91–92)

Development sustains itself – accelerates and extends itself – through its own internal dynamic alone (Lyotard 2004: 7). The end time brought about by the heat-death of the sun is used as a motif to represent the way that capitalist expansion refuses to be limited in any way – that it will attempt to overcome any end to its advance. This imperative to speed up is applied without concern for the human cost, without any concession to humanist projects or ideals. Capitalist development is an ideology

DOI: 10.1057/9781137394781.0006

that operates through saving time; it is blind to the costs of these savings. It transforms life itself into work, derives labour and value from mental energies, exploits our personalities and entraps our enthusiasm for communication. This parasite has colonised subjectivity itself and our lives are valued only in terms of their performative operation. We are left, as Lyotard argues, with an inhuman demand: 'be operational... or disappear' (Lyotard 2005: xxiv). We can understand this notion of the inhuman in two overlapping ways: the inhumanity of a capitalist system that advances whilst humans suffer; and the inhumanity that 'haunts the human from the inside' (Gane 2003: 439), taking the soul hostage (Lyotard 2004: 2). 'What else remains as "politics"', asks Lyotard (2004: 7), 'except resistance to this inhuman?' The focus of this final chapter is on the challenges to meeting this call for resistance.

Lyotard's concept of the inhuman is considered useful insofar as it locates the pathological development of communicational capitalism as a key site of struggle. His odd story demonstrates that amongst the limitations to development to be avoided are any concerns for the human consequences. Lyotard worried that subjection to capitalist principles of efficiency would begin to attack or diminish what it is to be human (see Hill 2012). He is concerned with what is lost when we move from a human to a computer mode of thinking. The human mind takes in the full picture, not neglecting side-effects or marginal data – focused but also lateral. Human thought is capable of discerning relevant information without fully processing all of the irrelevant information (Wallach 2010). Human thought is intuitive, hypothetical, it can work with ambiguous data and operate without pre-established rules or codes; it is 'a mode of thought not guided by rules for determining data, but showing itself as possibly capable of developing such rules afterwards on the basis of results obtained "reflexively" ' (Lyotard 2004: 15). This sort of reflexive judgement consists of 'the synthesis we are able to make of random data without the help of pre-established rules of linkage' (Lyotard 1988: 8). The computer mode, however, is limited to 'programming, forecasting, efficiency, security, computing, and the like'; it is 'the triumph of determinant judgement' (Lyotard 1988: 21). Those human qualities listed above are what allow us to be empathetic. We can quickly sift through all the communicational cues to reach an understanding of others, contextualise them within messy situations rather than subsume them under pre-given categories, intuit from fragmental utterances wider problems that people face, develop lines of exchange that are

DOI: 10.1057/9781137394781.0006

creative or unorthodox in order to reach deeper into the other's soul. This requires communication that has not been subjected to the performative demands of functionality and efficient transmission. But the sorts of communication that were considered in Chapter 3 were little more than the exchange of brute information, and something like the script of the call centre worker considered in Chapter 1 would neutralise the event of communication in advance. Our communication is stripped-down to the limitations of the technologies that facilitate it, reducing once meaningful exchange to sensitivity to information, to the pre-programmed and the forecasted. Human communication is pared-back 'to the immediate *processing* of information, and to the selection of pre-programmed, and thus standardised, options from the framework of the system' (Gane 2003: 441). This is disastrous for empathetic human communication. Corporations now have control over what is 'good' (efficient) communication and information, that is, what is relevant and what is not, what is operative and what is sub-optimal. Lyotard's concern is that whatever cannot be made efficient is eradicated, as 'those parts of the human race which appear superfluous' for the goal of continued development are 'abandoned' (Lyotard 2004: 77). Saving time is paramount; development is good in itself: there is no place for that old-fashioned idea of the good being that which is for the betterment of human society. Empathy has become inefficiency; the catastrophe we face is not so much solar as it is social. With this demand to replace empathetic communication with informational exchange, we risk losing the space within which solidarity can develop – our resistance is being curtailed by the medium of connectivity before we can even mobilise it.

What Lyotard has described as inhuman is essentially a kind of capitalist psychopathy, a system without empathy or remorse, without care or social anchorage, which moulds and constrains individuals into relations and modes of labour that are toxic, that are manipulative and irresponsible. Capital now turns a profit on the exploitation of life itself, of social essence and its spirit of communication, the soul put to work, and yet it has very little concern for life, and has, as demonstrated by the financial crisis of 2008, proven itself to be utterly toxic to society. Development has led us to precariousness and the brink of madness. Flexibility, which it was argued in Chapter 1 institutes a precariousness of time, is simply performativity or efficiency rebranded. Communication technologies constrain human thought into calculative, interpretive procedures, constituting a workplace that adopts what Franco Piperno (1996: 127)

DOI: 10.1057/9781137394781.0006

calls an 'operative culture', which drives itself by the demand to be ever faster, to optimise those procedures *ad infinitum*. Communication that is not operative, which does not maximally expand the efficiency of work, is obsolete for profit-making and so is eliminated. Communication is constrained into efficient modes, and workers speak according to rules informed by capitalist logic, which is as good as shutting people up altogether. And as we try to adapt to this demand for efficiency we undermine our own work conditions:

> Instead of profiting from the ease allowed by a production devoted to machines, humans find themselves competing against technology and are thus forced to reduce their demands and expectations to the level of machine. We try to work as much and as tirelessly as machines do, and by doing so we turn ourselves into second-rate production machines, never as efficient as the real ones. (Campagna 2013: 12)

As technologies develop, the domain of the Human Intelligence Task will contract, and we will have to sell our labour cheap to compete. Already we are working ourselves into nervous exhaustion to match the efficiency of communicative capitalism. At which point we face the choiceless choice of development: become efficient or become obsolete; medicate and carry on, or fall ill and fall behind. Psycho-pharmaceuticals will increase work performance and competitiveness amongst sick workers but production will be maintained only at great human cost. All of which demands resistance, as Lyotard would attest. However, his account of capitalist development as parasitical contains within it a barrier to struggle. Lyotard understands capitalist development as having run free of its tracks, no longer driven by human needs but propagating itself through human thought and creativity. Lyotard argues that the parasite will find a new host, artificial intelligence, once humans have run out of their Earthly resource. But more pressing is what happens when the parasite exhausts the mental energies that sustain it. The general intellect allows for cognitive labour to be extracted on a global scale. At the local level workers may be pushed to the brink of mental collapse. They then either succumb to see labour extracted from elsewhere; or medicate and try to out-compete alternative sources of cognitive labour. Who is the best bet for resistance, then: the burn-out or the junkie?

The biggest mental health problem we face today is debt. For the debtor it occupies all present thinking, every action constrained by this

DOI: 10.1057/9781137394781.0006

wholesale sell-off of individual futures. To be in debt is to be forced to anxiously apply the performative principle to every mundane aspect of our lives. This is an exercise of power over the horizon of people's possible life choices. Debt is a form of governance by guilt – Franco Berardi (2015: 26), perhaps more starkly, refers to it simply as black-mail – such that debtors are divested of autonomy and forced into miserable and precarious jobs to service what they owe. Bare existence becomes economic calculation, as we are forced to make trade-offs on life's essentials. *How can I most efficiently live to service my debts?* At its heart such a question represents the profound sadness of debt. Maurizio Lazzarato (2011: 55–56) reminds us that for Marx, debt was the most dehumanising form of estrangement of man from man because what is at stake is our social and moral existence – trust. The debtor's actions, their social existence and behaviour, are for the creditor measured only by the debtor's servicing of the debt; the value of human life is meas-ured by economic reason, such that the death of the debtor is only a tragedy to the creditor insofar as it is the death of their capital: 'Credit, then, not only exploits social relationships in general, but also the uniqueness of existence' (Lazzarato 2011: 59–60). But this idea that debt is dehumanising draws attention away from the real problem, which is that the whole system is inhuman; we do not, through debt, lose our humanity – it is co-opted into maintenance of efficient development. Life itself – our social and moral existence – is exploited by financial capitalism. So, whilst Lazzarato (2011: 73) argues that debt is not a pathology of capitalism, but rather one of the strategies adopted in the 1960s and 1970s to destroy the old capitalism and bring in the new; it is more accurate to emphasise that the development of the system is pathological and that the debt it creates constitutes a sickness spread-ing through society. Debt may not be considered a malady in financial capitalism, instead being part of its smooth functioning, but that is not to say that its operation is not sick; our present socioeconomic system is like some Bateman-esque psychopath, knowingly harming others as it goes about its daily business. Christian Marazzi (2011b: 96) describes the violent ethos of financial capitalism in terms that Lyotard would find familiar: 'time is everything, man is nothing'. Where debt can be a catastrophe for the individual but a boon for capital, there we see the inhumanity of a system that moves according to its own internal logic, with no concern for suffering or morality, driven by economic reason alone. 'Financial capitalism has achieved autonomy from social life',

DOI: 10.1057/9781137394781.0006

writes Berardi (2011: 140); we are in debt, development is saving (time), and the whole system is oblivious to the human cost. This saving of time constitutes the strongest relationship between melancholy and debt. Capital exerts control over the future in the form of debt, affecting future behaviour through guilt; we orient ourselves from present to future through our responsibility for money owed (Lazzarato 2011: 45–46): capitalist development is oriented towards possessing the future in advance – and we risk losing our hopes and dreams of a better tomorrow.

Berardi (2012: 8) argues that the financial crisis begun in 2008 is also 'a crisis of imagination about the future'. So many have seen their futures seized by indebtedness, that the present has become a project of working off debt, the radical horizon nothing more revolutionary than a clean credit rating. It is tempting to conclude, forlornly, that a different future is now withheld from us, sold off by a short-termist financial system in the form of credit, and that the potential for change will go unrealised. Berardi expresses this foreclosure well:

> I don't like the empty words of self-reassurance, or rhetoric about the multitude. I prefer to tell the truth, at least, the limited truth as I see it: there is no way out, social civilization is over, the neoliberal precarization of labour and the media dictatorship have destroyed the cultural antibodies that, in the past, made resistance possible. As far as I know. (2011: 158)

Such a statement is not as bleak as it appears on the surface. *As far as I know*: any radical event – such as whatever could rid us of the sadness of work and the pathology of communicative capitalism – is unpredictable; it is only planning for a different future that is impossible, not a different future to that planned and programmed, forecasted and foreclosed in the interests of capitalist development. This is perhaps where the possibility of resistance lies: we cannot know that our negative situation can be overcome because the event lies always beyond what we currently know. For Jacques Derrida (2007: 441–42), 'an event...must never be something that is predicted or planned, or ever really decided upon'; for Alain Badiou (2010: 222) 'an event *has, as a maximally true consequence of its (maximal) intensity of existence, the existence of an inexistent*', which is to say, the event creates a new possible (rather than realising something that is already possible); and for Lyotard (2007: 79), an event is when 'something happens which is not tautological with what has happened'. So there is hope – but it is impossible to say with any certainty what may

DOI: 10.1057/9781137394781.0006

be achieved with it. We have no future – and if we use this as a starting point then we may be able to think beyond communicative capitalism. As Berardi (2011: 163) counsels: 'The present ignorance has to be seen as the space of possibility'.

Perhaps the most fruitful space of possibility can be found in *melancholy*. At first glance this might appear to be a defeatist position but the important move here is to reposition our orientation to the present. Walter Benjamin (1974) first used the term 'left-wing melancholy' in his critique of the poetry of Erich Kästner, in which he locates an attachment to left ideas and analysis that is stronger than any inclination to transform the present with them. Wendy Brown (1999: 20) describes Benjamin's left melancholy as an 'unambivalent epithet for the revolutionary hack who is, finally, attached more to a particular political analysis or ideal – even to the failure of that ideal – than to seizing possibilities for radical change in the present'. The left melancholic clings to, but wallows in, nostalgia for things past – really existing socialism, an international left community, strong trade unions and workers' movements – in a way that renders them a conservative force in history (Brown 1999: 25). They are stuck in what Benjamin (1974: 29) identifies as the 'heaviness of heart' that comes with routine, a going-through-the motions of left politics without any motoricity, forlorn and ultimately purposeless. Jodi Dean (2012) updates this account, arguing that the left today enjoys its own powerlessness; so many are trapped in the circuits of communicative capitalism but cannot escape because they enjoy being there. On the one hand, the left is enticed by communication technologies, seeing a tool for political projects whilst conveniently ignoring the expropriation and the wasteful sluice of political energies:

> We talk, complain, and protest. We make groups on Facebook. We sign petitions and forward them to everyone in our contacts list. Activity becomes passivity, our stuckness in a circuit, which is then mourned as the absence of ideas or even the loss of the political itself and then routed yet again through a plea for democracy, although it doesn't take a genius to know that the real problem is capitalism. (Dean 2012: 66)

Somehow we have mistaken the tools of pathological development for the means of radical change, missing that something like the World Wide Web is really just a carpet sample for communicative capitalism and its principle of performativity. Dean (2009: 4) argues that the left

DOI: 10.1057/9781137394781.0006

enjoys the 'imaginary freedoms of creativity and transformation' that communication technologies provide, at the same time assuming – and even enjoying – the values of neoliberalism: individualism, circulation, saving time – all at the cost of solidarity. Jonathan Crary (2013: 121) supports this view with stark imagery, suggesting that the left's focus on online strategies and organisation is essentially a form of voluntary 'kettling' (the process whereby the police contain political demonstrations, not allowing protesters to leave an area for hours on end) 'where state surveillance, sabotage, and manipulation are far easier than in lived communities and localities where actual encounters occur'. On the other hand, there is something enjoyable about having been divested of the ability to effect change, as if the left is morally exculpated by the conditions of its own weakness – with a heavy heart, of course. This might best be understood with reference to Mark Fisher's (2009) concept of the bureaucratic libido. Fisher (2009: 49) defines this as 'the enjoyment that certain officials derive from [a] position of disavowed responsibility', that is, when a person can avoid human empathy or moral culpability with reference to inhuman bureaucratic structures or regulations – that they prop up with their own labour – and seemingly relish the authority with which such structures or regulations remove them from the moral universe. This is a kind of institutionalised enjoyment of the inflexibility of the system: 'it's not me, I'm afraid', as Fisher (2009: 49) characterises it, 'it's the regulations'. The bureaucratic libido is complicity without responsibility. The left melancholic libido, then, is an enjoyment of an inhuman system that demands we become operational or else obsolete, a meek acceptance of the terror of this ultimatum, and relief at the weakened position that removes the burden of resisting this inhumanity: 'it's not the left, I'm afraid; what could we do?' Communicative capitalist development is sustained by our stuckness in its circuits, our labour within the system, and to disavow any responsibility may even come as a relief. This kind of left melancholy must be resisted, not because it has given up on the future – but because it has given up on the present. This is not to say that melancholy is without use. Indeed, as Fisher (2009: 54) argues, to be happy under the conditions of communicative capitalism is something that can only be maintained 'if one has a near-total absence of any critical reflexivity'. The left melancholy identified by Benjamin, Brown and Dean is matched in cynicism only by those who can maintain a cheery outlook amidst all this misery. The most important project for the left is not to dwell on its own melancholic state but to repurpose

DOI: 10.1057/9781137394781.0006

and utilise the depressions and anxieties of a society made precarious by pathological development. As Fisher advises:

> We must convert widespread mental health problems from medicalized conditions into effective antagonisms. Affective disorders are a form of captured discontent; this disaffection can and must be channelled outwards, directed towards its real cause, Capital. (2009: 80)

There must be a way to grow solidarity out of vulnerability, strength out of precariousness, activism out of exhaustion and a new, healthy political present out of the psychopathology wrought by communicative capitalism. Melancholy now operates as a uniting force across society. It must become a site of struggle for a new solidarity, where depression and exploitation can be transformed into a refusal to submit to inhuman demands. That would involve rejecting utopian projects held in abeyance for what may be an endless struggle against inhumanity in the present, a struggle for autonomy, freedom from the sadness of work and its patho-logical re-organisation of society. We need to make the impossible real, and we need to do it now.

In series three of the BBC spy drama *Spooks,* MI5 operative Tom Quinn, played by Matthew Macfadyen, suffers what his boss calls a 'conscience explosion'. Throughout the series, Quinn's job sees his personality put to work to manipulate others, both plumbing the depths of his own identity to draw out and accentuate the appropriate characteristics to coerce civilians into defence of the realm and extract intelligence from those that would bring terror to it, and adopting new personas to infiltrate organisations and institutions, both friendly and threatening. His subjectivity is so constrained and manipulated into the contours of his role that he barely seems to know who he is. The man is a blank; his personality has been used up by the service. Quinn's dedicated gathering of information in order to counter terrorism eventually leads to mental instability. His social life is constantly eroded by work until it becomes non-existent. His attempts at a stable family are undermined by the exhaustion of his being, his *never not on* job requirements, and the refusal of his loved ones to accept the collateral damage of the world he moves in. Quinn's breakdown comes when he is ordered to ruin the life of yet another innocent, to be used as bait to entrap terrorists. The man's finances are wiped out, his reputation left in tatters and his family driven from him. At this point Quinn's conscience belatedly kicks in, he repudiates the inhumanity of the system he operates in, the moral murk

DOI: 10.1057/9781137394781.0006

of a network of spies that spread so much misery in the name of security, and attempts to extricate the patsy – his patsy – from the whole sorry affair. Quinn is immediately dismissed. Throughout this book it has been argued that cognitive labour is precarious labour, that communicative disease thrives in such conditions, that the organisation of our communication towards productive ends is dis-empathising and simultaneously plays to and reinforces social anxieties, that technological development is a parasite on the subjectivity of the worker, and that communicative capitalism is inherently pathological. The beginning point for resistance is the recognition that since cognitive labour is now the primary productive force in society, and that worker subjectivity is today generative of value, precariousness is something that afflicts the majority; that this primacy means that exhaustion is shared by all, and that the stresses and strains of work should never be accepted as part of the job description but as the source of a new unity; that instrumental communication is fragmenting society at the same time that we need renewed solidarity; and that there can be no acquiescence into enjoyment of a system whose operative principle brings misery and its own entrenchment without any concern for human betterment or the social costs. The left needs to mobilise a conscience explosion at both the individual level and that of the collective if it is to achieve autonomy over automation, freedom from the sadness of work, and to resist the inhumanity of communicative capitalism. Unless we build on our shared melancholy the future will remain a thing of the past. We cannot afford to be dismissed.

DOI: 10.1057/9781137394781.0006

References

Alba, D. 2015. 'Facebook at Work' Launches So You Can Never Not Be On Facebook. *Wired 14 January* http://www.wired.com/2015/01/facebook-at-work-launch.

Amazon 2014. https://www.mturk.com/mturk/welcome.

Badiou, A. 2010. *The Communist Hypothesis.* Trans. D. Macey and S. Corcoran. London: Verso.

Badiou, A. and S. Žižek. 2009. *Philosophy in the Present.* Ed. P. Engelmann. Trans. P. Thomas and A. Toscano. Cambridge: Polity.

Bauman, Z. 2004. *Work, Consumerism and the New Poor.* Buckingham: Open University Press.

Bauman, Z. 2008. *Liquid Modernity.* Cambridge: Polity.

Bauman, Z. and L. Donskis. 2013. *Moral Blindness: The Loss of Sensitivity in Liquid Modernity.* Cambridge: Polity.

Baudrillard, J. 1995. *The Gulf War Did Not Take Place.* Trans. P. Patton. Bloomington, IN: Indiana University Press.

Baudrillard, J. 2009. *The Transparency of Evil: Essays on Extreme Phenomena.* Trans. J. Benedict. London: Verso.

Baudrillard, J. 2011. *Impossible Exchange.* Trans. C. Turner. London: Verso.

Benjamin, W. 1974. Left-Wing Melancholy (On Erich Kästner's New Book of Poems). *Screen,* 15:2, 28–32.

Berardi, F. 2007. Anatomy of Autonomy. In *Autonomia: Post-Political Politics,* 148–70. Eds. S. Lotringer and C. Marazzi. Los Angeles, CA: Semiotext(e).

Berardi, F. 2009a. *Precarious Rhapsody: Semiocapitalism and the Pathologies of the Post-Alpha Generation.* Eds. E. Empson and S. Shukaitis. Trans. A. Bove *et al.* London: Minor Compositions.

Berardi, F. 2009b. *The Soul at Work.* Trans. F. Cadel & G. Mecchia. Los Angeles, CA: Semiotext(e).

Berardi, F. 2011. *After the Future.* Eds. G. Genosko and N. Thoburn. Trans. A. Bove *et al.* Edinburgh: AK Press.

Berardi, F. 2012. *The Uprising: On Poetry and Finance.* Los Angeles, CA: Semiotext(e).

Berardi, F. 2015. *Heroes: Mass Murder and Suicide.* London: Verso.

Boden, D. and H. Molotch. 2004. Cyberspace Meets the Compulsion of Proximity. In *The Cybercities Reader,* 101–105. Eds. S. Graham. London: Routledge.

Brown, W. 1999. Left Melancholy. *Boundary 2,* 26:3, 19–27.

Burke, E. 2008. *A Philosophical Enquiry into the Origin of our Ideas of the Sublime and Beautiful.* Ed. A. Phillips. Oxford: Oxford University Press.

Campagna, F. 2011. Recurring Dreams: The Red Heart of Fascism. *Through Europe* http://th-rough.eu/writers/campagna-eng/recurring-dreams-red-heart-fascism.

Campagna, F. 2013. *The Last Night: Anti-Work, Atheism, Adventure.* Winchester: Zero Books.

Crary, J. 2013. *24/7: Late Capitalism and the Ends of Sleep.* London: Verso.

Dean, J. 2009. *Democracy and Other Neoliberal Fantasies: Communicative Capitalism and Left Politics.* London: Duke University Press.

Dean, J. 2010. *Blog Theory: Feedback and Capture in the Circuits of Drive.* Cambridge: Polity.

Dean, J. 2012. *The Communist Horizon.* London: Verso.

Deleuze, G. 1992. Postscript on the Societies of Control. *October,* 59 (Winter), 3–7.

Derrida, J. 2007. A Certain Impossibility of Saying the Event. *Critical Inquiry,* 33:2, 441–61.

Eagleton, T. 2009. *Trouble with Strangers: A Study in Ethics.* Chichester: Wiley-Blackwell.

Fisher, M. 2009. *Capitalist Realism: Is There No Alternative?* Winchester: Zero Books.

Fisher, M. 2011. The Future is Still Ours: Autonomy & Post-Capitalism. In *We Have Our Own Concept of Time and Motion* (part one), 5–7.

Gane, N. 2003. Computerised Capitalism: The Media Theory of Jean-François Lyotard. *Information, Communication & Society,* 6:3, 430–50.

Gill, R. and A. Pratt. 2008. In the Social Factory? Immaterial Labour, Precariousness and Cultural Work. *Theory, Culture & Society,* 25:7–8, 1–30.

Graeber, D. 2013. On the Phenomenon of Bullshit Jobs. *Strike! Magazine* http://strikemag.org/bullshit-jobs/

Hardt, M. 1996. Laboratory Italy. In *Radical Thought in Italy: A Potential Politics*, 1–10. Eds. P. Virno and M. Hardt. Minneapolis, MN: University of Minnesota Press.

Harvey, D. 2012. *Rebel Cities: From the Right to the City to the Urban Revolution*. London: Verso.

Hill, D. W. 2012. Jean-François Lyotard and the Inhumanity of Internet Surveillance. In *Internet and Surveillance: The Challenges of Web 2.0 and Social Media*, 106–23. Eds. C. Fuchs *et al.* London: Routledge.

Humphry, J. 2014. Visualising the Future of Work: Myth, Media and Mobilities. *Media, Culture & Society*, 36:3, 351–66.

Jensen, T. 2012. Tough Love in Tough Times. *Studies in the Maternal*, 4:2 http://www.mamsie.bbk.ac.uk/back_issues/4_2/documents/ Jensen_SiM_4(2)2012.pdf.

Jones, O. 2012. *Chavs: The Demonization of the Working Class*. London: Verso.

Lazzarato, M. 1996. Immaterial Labor. In *Radical Thought in Italy: A Potential Politics*, 133–47. Eds. P. Virno and M. Hardt. Minneapolis, MN: University of Minnesota Press.

Lazzarato, M. 2011. *The Making of the Indebted Man: An Essay on the Neoliberal Condition*. Trans. J. D. Jordan. Los Angeles, CA: Semiotext(e).

Lazzarato, M. 2014. *Signs and Machines: Capitalism and the Production of Subjectivity*. Trans. J. D. Jordan. Los Angeles, CA: Semiotext(e).

Lewis, J. and A. West. 2009. 'Friending': London-Based Undergraduates' Experience of Facebook. *New Media & Society*, 11:7, 1209–29.

Lotringer, S. 2007. In the Shadow of the Red Brigades. In *Autonomia: Post-Political Politics*, v–xvi. Eds. S. Lotringer and C. Marazzi. Los Angeles, CA: Semiotext(e).

Lyotard, J.-F. 1988. *Peregrinations: Law, Form, Event*. Guildford: Columbia University Press.

Lyotard, J.-F. 1992. *The Postmodern Explained to Children: Correspondence 1982-1985*. Trans. D. Barry et al. Eds. J. Pefanis and M. Thomas. London: Turnaround.

Lyotard, J.-F. 1993. *Toward the Postmodern*. Trans. K. Berri *et al.* Eds. R. Harvey and M. S. Roberts. Atlantic Highlands, NJ: Humanity Books.

Lyotard, J-F. 2004. *The Inhuman: Reflections on Time*. Trans. G. Bennington and R. Bowlby. Cambridge: Polity Press.

DOI: 10.1057/9781137394781.0007

Lyotard, J-F. 2005. *The Postmodern Condition: A Report on Knowledge.* Trans. G. Bennington and B. Massumi. Manchester: Manchester University Press.

Lyotard, J-F. 2007. *The Differend: Phrases in Dispute.* Trans. G. Van Den Abbeele. Minneapolis, MN: University of Minnesota Press.

Marazzi, C. 2008. *Capital and Language: From the New Economy to the War Economy.* Trans. G. Conti. Los Angeles, CA: Semiotext(e).

Marazzi, C. 2011a. *Capital and Affects: The Politics of the Language Economy.* Trans. G. Mecchia. Los Angeles, CA: Semiotext(e).

Marazzi, C. 2011b. *The Violence of Financial Capitalism.* Trans. K. Lebedeva and J. F. McGimsey. Los Angeles, CA: Semiotext(e).

Marsters, K. 2014. Linking Mental Health Treatment to Job Support is a Cruel Concept. *Guardian 6 August* http://www.theguardian.com/commentisfree/2014/aug/06/mental-health-treatment-job-support-benefits.

Marx, K. 1993. *Grundrisse: Foundations of the Critique of Political Economy.* Trans. M. Nicolaus. London: Penguin.

Mitropoulos, A. 2006. Precari-Us. *Mute 9 January* http://www.metamute.org/editorial/articles/precari-us.

Moulier Boutang, Y. 2011. *Cognitive Capitalism.* Trans. E. Emery. Cambridge: Polity.

Naughton, J. 2015. What Can We Learn from Facebook's Annual Bullshit Report? *Guardian 25 January* http://www.theguardian.com/technology/2015/jan/25/facebook-annual-bullshit-report.

Pansa, G. 2007. Fiat Has Branded Me. In *Autonomia: Post-Political Politics,* 24–27. Eds. S. Lotringer and C. Marazzi. Los Angeles, CA: Semiotext(e).

Piperno, F. 1996. Technological Innovation and Sentimental Education. In *Radical Thought in Italy: A Potential Politics,* 123–30. Eds. P. Virno and M. Hardt. Minneapolis, MN: University of Minnesota Press.

Revelli, M. 1996. Worker Identity in the Factory Desert. In *Radical Thought in Italy: A Potential Politics,* 115–20. Eds. P. Virno and M. Hardt. Minneapolis, MN: University of Minnesota Press.

Russell, B. 2004. *In Praise of Idleness.* London: Routledge.

Savage, M. *et al.* 2013. A New Model of Social Class? Findings from the BBC's Great British Class Survey Experiment. *Sociology,* 47:2, 219–50.

Sennett, R. 1999. *The Corrosion of Character: The Personal Consequences of Work in the New Capitalism.* London: W. W. Norton and Co.

Sennett, R. 2012. *Together: The Rituals, Pleasures and Politics of Co-operation.* London: Allen Lane.

DOI: 10.1057/9781137394781.0007

Seymour, R. 2012. We Are All Precarious: On the Concept of the Precariat and its Misuses. *New Left Project 10 February* http://www.newleftproject.org/index.php/site/article_comments/we_are_all_precarious_on_the_concept_of_the_precariat_and_its_misuses.

Smith, A. 2009. *The Theory of Moral Sentiments.* Ed. R. P. Hanley. London: Penguin.

Standing, G. 2013. *The Precariat: The Dangerous New Class.* London: Bloomsbury.

Stiegler, B. 2011. *For a New Critique of Political Economy.* Trans. D. Ross. Cambridge: Polity.

Terranova, T. 2004. *Network Culture: Politics for the Information Age.* London: Pluto Press.

Terranova, T. 2012. Attention, Economy and the Brain. *Culture Machine,* 13 http://www.culturemachine.net/index.php/cm/article/view/465/484.

Turkle, S. 2011. The Tethered Self: Technology Reinvents Intimacy and Solitude. *Continuing Higher Education Review,* 75, 28–31.

Turkle, S. and J. Nolan. 2012. A Conversation with Sherry Turkle. *The Hedgehog Review,* 14:1, 53–64.

Virilio, P. 1998. *Open Sky.* Trans. J. Rose. London: Verso.

Virilio, P. 2012. *The Administration of Fear.* Trans. A. Hodges. Los Angeles, CA: Semiotext(e).

Virilio, P. and F. Kittler. 2001. The Information Bomb: A Conversation. In *Virilio Live: Selected Interviews,* 97–109. Ed. J. Armitage. London: Sage Publications.

Virno, P. 1996. The Ambivalence of Disenchantment. In *Radical Thought in Italy: A Potential Politics,* 13–34. Eds. P. Virno and M. Hardt. Minneapolis, MN: University of Minnesota Press.

Virno, P. and M. Hardt (eds.) 1996. *Radical Thought in Italy: A Potential Politics.* Minneapolis, MN: University of Minnesota Press.

Wainwright, O. 2012. Like-A-Hug? The Facebook Vest that Gives You a Hug from Your Friends. *Guardian 9 October* http://www.theguardian.com/culture/2012/oct/09/like-a-hug-facebook-vest-friends

Wallach, W. 2010. Robot Minds and Human Ethics: The Need for a Comprehensive Model of Moral Decision Making. *Ethics and Information Technology,* 12:3, 243–50.

Woodcock, J. 2014. Precarious Workers in London: New Forms of Organisation and the City. *City,* 18:6, 776–88.

DOI: 10.1057/9781137394781.0007

Wright, S. 2002. *Storming Heaven: Class Composition and Struggle in Italian Autonomist Marxism.* London: Pluto Press.

YouTube 2012. Amazon New Kindle TV Commercial https://www.youtube.com/watch?v=2EQoe7dYuaI

Žižek, S. 1999. Is it Possible to Traverse the Fantasy in Cyberspace? In *The Žižek Reader,* 102–24. Eds. E. Wright and E. Wright. Oxford: Blackwell.

Žižek, S. 2008. *Violence: Six Sideways Reflections.* London: Profile Books.

Žižek, S. 2010. *Living in the End Times.* London: Verso.

DOI: 10.1057/9781137394781.0007

Index

ADHD, 34, 37
alternative officing, 50–51
Amazon, 19, 42
anxiety, 36, 38, 42, 45, 47–8,
    48, 54
attention, 32–5, 37–8
automation, 3–4, 10–11, 16, 18,
    36, 48–9, 66
autonomy, 3–4, 10–11, 14, 16,
    23–5, 31, 49, 51, 61, 65–6

Badiou, A., 62
Baudrillard, J., 39, 40–2, 47–8
Bauman, Z., 24, 36, 45, 51, 53
Benjamin, W., 63, 64
Berardi, F., 3–4, 6, 7, 10, 16,
    17–18, 19, 21, 24, 26, 27, 31–2,
    34, 36, 37, 46, 51, 53, 61–3
bullshit, 8, 53

call centre, 12–15, 30
Campagna, F., 20–1, 38, 60
Cognitive Behavioural
    Therapy, 38–9
Crary, J., 26, 34, 35, 44, 50, 64

Dean, J., 5, 7, 8–9, 32, 44, 47, 48,
    53, 63
debt, 6, 38, 60–2
Deleuze, G., 14
depression, 37, 38, 64–5
Derrida, J., 62
development, 3–6, 10–11, 25–7,
    33–4, 39, 55–8, 66

empathy, 16, 49–50, 51, 53, 54,
    58–9, 64, 66

Facebook, 1–3, 44, 45–6,
    46, 53
Fisher, M., 4, 6, 15, 34, 37,
    64–5
flexibility, 3–4, 6, 9, 13, 18–20,
    20–1, 23–4, 27–8, 35, 36,
    48–9, 50–1, 59, 64
future, 3–4, 6, 7–8, 36, 60–3,
    64, 66

general intellect, 18–19, 21, 28,
    31–2, 34, 35–6, 60
Google, 7

Hardt, M., 10, 20

inhuman, 6, 10, 55–62, 64–6
insomnia, 30, 35, 38
interface simulacrum, 44–5

Lazzarato, M., 6, 7, 14, 17, 24,
    25, 52, 60–2
left melancholy, 63
Lyotard, J.-F., 4, 8, 55–9, 62

Marazzi, C., 4, 15–16, 17, 18, 20,
    23, 27, 31, 33, 61
Marx, K., 18, 61
melancholic libido, 64–5
Moulier Boutang, Y., 7, 15, 19,
    27, 33

precariousness, 3, 5, 7, 9–11, 13, 19–24, 27–8, 30, 31, 33, 35, 36, 38, 48, 49, 51, 53, 59, 64–6
propaganda, 4–6
psycho-pharmaceuticals, 37–8, 39, 60, 65

Sennett, R., 22, 24, 25, 36, 50, 51, 53
stalking, 45–6
Standing, G., 9, 21–4, 35, 38, 48, 51
Stiegler, B., 16, 26
stress, 31–2, 34

Tamagotchi, 44–5
Terranova, T., 33, 54
Twitter, 8, 53

Virilio, P., 5, 9, 33, 34, 42–3, 47
Virno, P., 13, 16–17, 24, 36

work time, 1–3, 4, 5, 6, 7, 10, 13, 19, 24–7, 32–3, 35, 65

Žižek, S., 44–5, 45–6

DOI: 10.1057/9781137394781.0008